Ying Chen

Deformation Behavior of Thin Metallic Wires under Tensile and Torsional Loadings

**Schriftenreihe
des Instituts für Angewandte Materialien**
Band 26

Karlsruher Institut für Technologie (KIT)
Institut für Angewandte Materialien (IAM)

Eine Übersicht über alle bisher in dieser Schriftenreihe erschienenen Bände
finden Sie am Ende des Buches.

Deformation Behavior of Thin Metallic Wires under Tensile and Torsional Loadings

by
Ying Chen

Dissertation, Karlsruher Institut für Technologie (KIT)
Fakultät für Maschinenbau
Tag der mündlichen Prüfung: 07. März 2013

Impressum

Karlsruher Institut für Technologie (KIT)
KIT Scientific Publishing
Straße am Forum 2
D-76131 Karlsruhe
www.ksp.kit.edu

KIT – Universität des Landes Baden-Württemberg und
nationales Forschungszentrum in der Helmholtz-Gemeinschaft

KIT Scientific Publishing 2013
Print on Demand

ISSN 2192-9963
ISBN 978-3-7315-0049-0

Deformation Behavior of Thin Metallic Wires under Tensile and Torsional Loadings

Zur Erlangung des akademischen Grades

Doktor der Ingenieurwissenschaften

der Fakultät für Maschinenbau

Karlsruher Institut für Technologie (KIT)

genehmigte

Dissertation

von

M. Sc. Ying Chen

aus Jiangsu

Tag der mündlichen Prüfung: 07.03. 2013
Hauptreferent: Prof. Dr. Oliver Kraft
Korreferent: Prof. Dr. Peter Gumbsch

Abstract

Size effects are widely observed in the mechanics of materials at the micron scale. However, the underlying deformation mechanisms remain ambiguous, particularly in the presence of strain gradients. In this work, combined microstructural investigations and mechanical tests (tension and torsion) were conducted on polycrystalline gold micro wires with different diameters (D = 15, 25, 40 and 60 μm) to determine the influences of specimen size, grain size, strain rate and loading type on the deformation behavior of the wires. The as-received, intermediate and fully-recrystallized states are distinguished based on its respective microstructure and mechanical behavior. It was found, that the strain rate does not significantly affect both the tensile and torsional strength. This is hypothesized to be related to the grain size of the wires. Hall-Petch effect was observed in the fully recrystallized wires under tensile and torsional loadings. However, the Hall-Petch effect in tension shows a specimen size dependency, whereas it is size independent in torsion. This difference can be traced back on the directionality of dislocation movements in dependence of the loading mode. It is argued that dislocations may leave the grains at the free surface during tensile deformation. However, the dislocations are driven towards the center of wires in torsion depending on the additional strain gradient. Moreover, continuous size effects were observed for different thick wires under torsional loadings at various twist rates, which was validated by the fully recrystallized 12.5 and 17.5 μm wires. The size effects were found to be a concurrent outcome from different degrees of pre-deformation, Hall-Petch effect, texture differences and the occurrence of strain gradients. Furthermore, the distribution of geometrically necessary dislocations (GNDs), determined from EBSD investigations, was found to be affected by both strain gradients and grain boundaries. The grain boundaries play a dominant role at small strains in fine grained structures, while the influence of strain gradients becomes more significant at large strains in coarse grained structures. With increasing shear strain, the highest GND densities were found evolving from the surface to the center of the wires. Besides, some additional investigations on polycrystalline aluminum micro wires (D = 15, 17.5, 25, 40 µm) were performed to ascertain whether metals with the same crystal lattice show similar characteristics in small dimensions under comparable conditions. Strain rate sensitivity on the proof stress of the aluminum wires was observed in both tension and torsion. Additionally, it was found that the uniform elongation ε_u of aluminum wires increases with increasing displacement rate. A size effect was observed for wires under torsional loadings.

Kurzfassung

Auf der Mikrometer-Skala werden im mechanischen Verhalten von Werkstoffen oftmals Größeneffekte beobachtet. Die zugrunde liegenden Verformungsmechanismen bleiben jedoch (häufig) unklar, besonders in der Gegenwart von Dehnungsgradienten. Innerhalb dieser Arbeit wurden kombinierte mikrostrukturelle Untersuchungen und mechanische Prüfungen (Zug und Torsion) an polykristallinen Gold-Mikrodrähten mit verschiedenen Durchmessern (D = 15, 25, 40 und 60 μm) durchgeführt, um den Einfluss der Probengröße, Korngröße, Dehnrate und Belastungsart auf das Verformungsverhalten der Drähte zu bestimmen. Die verschiedenen Zustände des Materials, das im Anlieferzustand, sowie im teilweise und vollständig rekristallisierten Zustand vorlag, wurden durch die jeweiligen Gefüge und das mechanische Verhalten unterschieden. Es wurde festgestellt, dass die Dehnrate keinen wesentlichen Einfluss sowohl auf die Zug- als auch auf die Torsionsfestigkeit hat. Dabei wird angenommen, dass dies auf die Korngröße der Drähte zurückzuführen ist. Ein Hall-Petch-Verhalten wurde für vollständig rekristallisierte Drähte jeweils unter Zug- und Torsionsbelastungen beobachtet. Allerdings zeigt der Hall-Petch Effekt eine Abhängigkeit von der Probengröße unter Zugbeanspruchung, während er unter Torsionsbeanspruchung unabhängig von der Probengröße ist. Dieser Unterschied kann auf die Richtungsabhängigkeit der Versetzungsbewegungen je nach Belastungsmodus zurückgeführt werden. Hierbei wird argumentiert, dass unter Zugbelastung Versetzungen die Körner an der freien Oberfläche verlassen können. Unter Torsionsbeanspruchung hingegen werden die Versetzungen, abhängig vom zusätzlichen Dehnungsgradienten, zur Mitte der Drähte hin getrieben. Außerdem wurden kontinuierliche Größeneffekte für die unterschiedlich dicken Drähte bei verschiedenen Dehnraten unter Torsionsbeanspruchung beobachtet, die durch zusätzliche Untersuchungen an vollständig rekristallisierten 12,5 und 17,5 µm starken Drähten verifiziert wurden. Die Größeneffekte erwiesen sich als zusammengesetzt aus mehreren Effekten, basierend auf unterschiedlichen Vorverformungsgraden, variierenden Hall-Petch-Verhalten, Texturunterschieden und dem Auftreten von Dehnungsgradienten. Darüber hinaus wurde anhand von EBSD Untersuchungen ermittelt, dass die Verteilung von geometrisch notwendigen Versetzungen (engl. geometrically necessary dislocations, kurz GND) sowohl von Dehnungsgradienten als auch von Korngrenzen beeinflusst wird. Die Korngrenzen spielen eine bedeutende Rolle bei kleinen Dehnungen in feinkörnigen Strukturen, während der Einfluss des Dehnungsgradienten eine stärkere Bedeutung bei großen Dehnungen in grobkörnigen Strukturen aufweist. Es wurde herausgefunden, dass sich mit zunehmender Scherdehnung die Position der höchsten GND-Dichte von der Oberfläche zur Mitte der Drähte hin verschiebt. Zusätzlich wurden einige Untersuchungen an polykristallinen Aluminium-Mikrodrähten (D = 15, 17,5, 25, 40 µm) durchgeführt um nachzuprüfen, ob Metalle mit gleichem Kristallgitter unter vergleichbaren Bedingungen ähnliche Eigenschaften in kleinen Dimensionen zeigen. Eine Dehnratenempfindlichkeit der Streckgrenze der Aluminiumdrähte wurde sowohl unter Zug- als auch unter Torsionbelastung beobachtet. Ferner wurde festgestellt, dass die Gleichmaßdehnung von Aluminiumdrähten mit zunehmender Verformungsrate zunimmt. Ein Größeneffekt wurde für Drähte unter Torsionsbeanspruchung festgestellt.

Acknowledgements

This work, supported by Helmholtz Association and China Scholarship Council, was carried out during my Ph.D. study at the Institute of Applied Materials (IAM-WBM) of Karlsruhe Institute of Technology from September 2009 to September 2012. I would like to express my sincere gratitude to all of them who supported me during this time.

First of all, I am extremely grateful to my research guide, Prof. Dr. Oliver Kraft, who always offered helpful advice and support. I would also like to thank Prof. Dr. Oliver Kraft for being my first examiner to give me the opportunity to obtain my doctorate at his institute.

I would like to thank Prof. Dr. Peter Gumbsch for being my second examiner and significant inputs to my dissertation. I want to acknowledge PD. Dr. Jarir Aktaa, as head of the department of Mechanics of Solids II, for his support and supervise during my work.

Sincere thanks go to my supervisor, Dr. Mario Walter, for his guidance. As my supervisor, he spent a lot of time to help my work so that I benefit hugely from his ideas and our discussions.

Besides, I would like to thank faculty members of our Institute, who created a nice work environment. They offered me not only the professional knowledge but also personal experience. I highly appreciate the great effort of Dr. RuiPing Hoo on my thesis and kind help from Dr. Widodo Widjaja Basuki.

Acknowledgements

Finally, I would like to thank my parents for their support and understanding on my study. I also thank all the friends in Germany or in China.

Karlsruhe,   Ying Chen
March, 2013

Table of Content

1. Introduction

Material properties change significantly with a reduction in dimension towards the micron scale. There is a general trend such that the smaller the characteristic length scale of a material the stronger the material. The influence of the size on plastic deformation, as well as possible changes of the deformation mechanisms in dependence of the size, have attracted tremendous attention in the last decades. This is because the observed mechanical responses of metallic materials in small dimensions cannot be predicted by the well-established classical non-linear continuum models. In this chapter, the background, the objectives and the scopes of this PhD thesis are described.

1.1 Background

With the increasing trend of miniaturizing devices, there are great demands for micro components in various applications, namely in optoelectronics, mass storage, medicine, biotechnology, communications, avionics etc.. In order to accomplish the requirements for such components under different operating conditions, appropriate materials have to be carefully chosen for every single application. Gold is in particular an important material, which is widely used in the field of micro electro-mechanical system (MEMS), due to its excellent electrical and thermal conductivity as well as high ductility and outstanding corrosion resistance [1]. So far, gold micro wires have a wide range of applications in MEMS packaging, such as electrical bonding of micro electro-mechanical system (MEMS) devices into integrated circuits for controlling and sensing. The use of gold in electronics industry has maintained a steady growth over the past two decades. Thus, it is necessary to investigate the mechanical behavior of gold wires under different loading conditions corresponding to possible loadings under operating conditions, such as bending, torsion and tension etc., in the presence as well as in the absence of strain gradients.

In the past, various gradient-depending theories [2-14] have been developed to account for the influence of the observed size effects on the deformation behavior of materials. However, the underlying deformation mechanisms remain ambiguous, particularly in the presence of strain gradients. Moreover, the influence of material parameters (e.g., grain size, specimen size, dislocation density etc.) and loading conditions (e.g., strain rate, monotonic / cyclic loading) on the deformation behavior in the presence of strain gradients in most cases remain unclear. Hence, appropriate experimental investigations have to be conducted in order to validate or to modify existing mechanical models.

1.2   Objectives and scopes of this work

This work aims to investigate the influence of microstructural properties (e.g., grain size, specimen size, texture etc.) and loading conditions (e.g., strain rate) on the deformation behavior of metallic materials in small dimensions in the presence of strain gradients. To achieve these goals, combined microstructural investigations and mechanical tests, more precisely uniaxial micro-tensile as well as multi-axial micro-torsion tests, were performed on polycrystalline gold micro wires with different diameters. In this study, comprehensive heat treatment investigations were done to obtain different microstructures of interest. The distributions of geometrically necessary dislocations, which are obviously related to the strain gradients, were analyzed by electron backscatter diffraction (EBSD).

Furthermore, additional investigations on polycrystalline aluminum micro wires were performed in order to ascertain whether metals with the same lattice show similar characteristics in small dimensions under comparable loading conditions.

In Chapter 2, the relevant state-of-art research for this work, concerning mechanical test methods, size effects, strain gradient theories and some superimposing effects on the mechanical behavior (e.g., texture, recovery/recrystallization processes, strain rate effects) is reviewed. Chapter 3 describes the experimental methods for the microstructural studies and mechanical characterizations. The results of the investigations are summarized in Chapter 4. Chapter 5 gives a detailed discussion on these obtained results. Finally, the conclusions of this work are summarized in Chapter 6.

2.　State of the art

In this chapter, the related mechanical testing methods and the fundamental knowledge on size effects are reviewed. In the first section, micro-tension and micro-torsion are introduced. The strain gradient plasticity theories based on geometrically necessary dislocations (GNDs) are reviewed in the second section. An overview of size effects found in experimental observations, categorized as intrinsic and extrinsic size effects, is given in the third section. Finally, the knowledge of the influence of texture and strain rate on the deformation behavior as well as the influence of recovery/ recrystallization processes occurring during heat treatments is briefly introduced in the last section.

2.1　Micro-tension and micro-torsion

Tensile test is most often used to determine a material's mechanical response under uniaxial loadings, while torsion test is an effective way to create strain gradients in a specimen with a high resolution at low-strain level under multi-axial loading [15]. Figures 2.1(a) and (b) show schematic diagrams of cylindrical specimens under both tensile and torsional loading. In tension, when a specimen is subjected to an external force F, it undergoes a change in shape with an increase in length from l_0 to $l_0+\Delta l$. The strain, the dimensionless normalized extension, is described as: $\varepsilon = \dfrac{\Delta l}{l_0}$.

In torsion, an applied twisting moment (M_t) results in a relative angular displacement of points along the circumference (and the interior) of a specimen. For specimens with circular cross-sections, each cross-section rotates in its own plane without warping. This kind warping-free torsion is called St. Venant torsion. In contrast, when the same torsional moment is applied to a noncircular cross-section of a bulk material, the cross-sections tend to warp in the z-direction, which is called warping torsion. Theoretically, the resistance to St. Venant torsion is provided only by shear stress in the cross-sectional planes, while warping torsion is carried by both shear and axial stresses. If warping is totally unrestrained, all cross-sections will only experience St. Venant torsion.

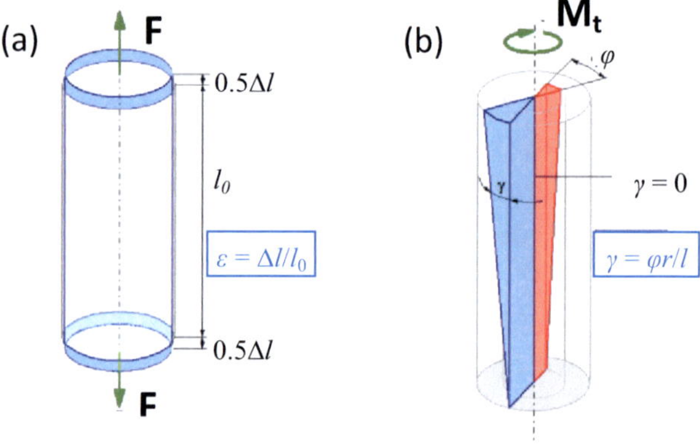

Figure 2.1: Schematic diagrams of cylinders under (a) tension and (b) torsion. In tension where strain is defined as $\varepsilon = \Delta l / l_0$, sample length l_0 increases Δl when subjected to external force F. In torsion where shear strain is described as $\gamma = \varphi r / l$, r is the radial position, φ is the displacement angle, and l is the length.

However, when end conditions and geometry restrain warping, the noncircular cross-sections will be subjected to the normal and shear stresses of warping torsion. For warping-free cross-sections, the shear strain γ varies linearly in the radial direction from the center ($\gamma = 0$) towards the surface ($\gamma = \gamma_{max}$) in dependence of the twist per length κ:

$$\gamma = \kappa \cdot r = \frac{\varphi}{l} \cdot r,$$

where r is the radial position, $\kappa = \varphi/l$ is the twist per unit length, φ is the displacement angle, and l is the length. However, based on the high precision requirement for the alignment of specimens into the rotation axis for the test set-up, micro-torsion experiments are rarely conducted and the number of published investigations is still relatively small [16-18].

A pioneer work on micro-tensile and micro-torsion experiments was conducted by Fleck and his coworker [16] two decades ago. Figure 2.2 shows the results of their investigations obtained from experiments on annealed copper wires of diameters ranging from 12 to 170 μm. A systematic trend in tension between the strength and the diameter was not found. In contrast, a significant size effect was observed in torsion, which shows "smaller is stronger". The torsional response was determined by the average shear stress (Q/a^3) over the cross-section as a function of shear strain at the surface of a wire (κa), where Q is the twisting moment, a is the wire radius and $\kappa = \varphi/l$ is the twist per unit length. Based on these observations, they introduced a phenomenological plasticity law whereby the stress is a function of both strain and strain gradient.

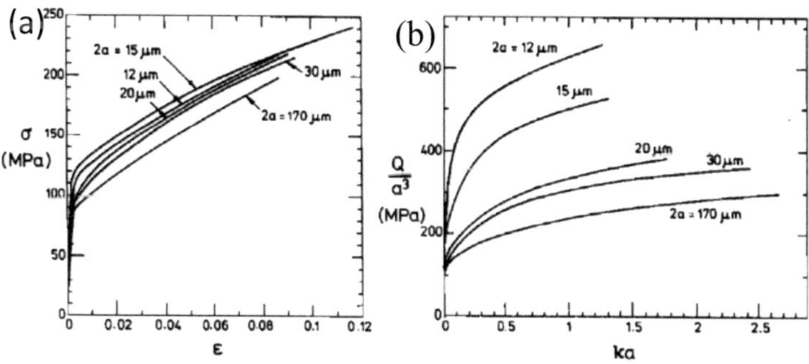

Figure 2.2: The mechanical response of copper wires under (a) tensile and (b) torsional loadings (Reproduced from Ref. [16]).

Although increasing grain sizes (ranging from 5 μm to 25 μm) with increasing wire diameters were identified, microstructural influences on the size effect were not taken into account in this work. Two other research groups further investigated copper wires in micro-torsion tests. Dunstan and co-workers [17] tested copper wires with diameters of 10 μm and 50 μm using a load-unload technique. A size effect was reported in the strength of wires for both initial yielding and flow stress under torsional loadings. Liu et al. [18] conducted micro-torsion experiments on annealed copper wires with diameters ranging from 18 μm to 105 μm using a new automated torsion balance technique. They observed a significant size effect in torsion, whereas only an insignificant size effect was found in tension. The influence of the grain size was considered in these further investigations but it was not quantitatively discussed since the influence of grain size was assumed to be small compared to the impact of strain gradients.

2.2 Strain gradient theories

The size dependency of material behavior observed in various experiments cannot be predicted by the classical continuum theories since the constitutive models do not have internal length scale component. In all the experiments, where strain gradients were involved, there is a general agreement that the size effect can be attributed to hardening caused by geometrically necessary dislocations [19, 20].

A phenomenological plasticity law was developed by Fleck et al. to account for the size effect observed in torsion, based on the strain gradient which exists between the surface and the center of the wire. The plastic strain gradient (η_{pl}) can generally be expressed as [21]:

$$\eta_{pl} = \frac{\varepsilon_{pl}}{l} \qquad (2.1)$$

where ε_{pl} is plastic strain, l is the dimensional length scale, in the case of torsion l is the wire diameter r.

The basis of this strain gradient theory is the assumption that hardening arises from the accumulation of both randomly stored and geometrically necessary dislocations [16, 19, 20]. When a material is plastically deformed, there are two types of dislocations stored, which are the "statistically stored dislocations (SSDs)" and the "geometrically necessary dislocations (GNDs)" [20]. Statistically stored dislocations trap each other randomly, which is associated with the uniform deformation [20]. When materials are plastically and non-homogeneously deformed, local strain gradients are present and geometrically necessary dislocations are additionally stored in order to accommodate the lattice curvature, as shown in Figure 2.3. Figure 2.3(b) depicts the deformation that each grain would show in the absence of constraints, assuming each grain is deformed as a single crystal. In this case, voids or grain "overlap" would ensue, as shown in Figure 2.3(c). However, the voids or overlaps do not appear in reality, otherwise the material would fail. It is proposed that the geometrically necessary dislocations arising from the local strain gradients at the grain boundaries accommodate the lattice curvature and eliminate the voids or overlaps, as illustrated in Figures 2.3(d).

The density of geometrically necessary dislocations is directly related to the plastic strain gradient (η_{pl}) [22]:

$$\rho_{gnd} = \frac{\eta_{pl}}{b} \qquad (2.2)$$

where b is the magnitude of the Burgers vector. In the case of torsion, η_{pl} can be interpreted as the twist per unit length κ.

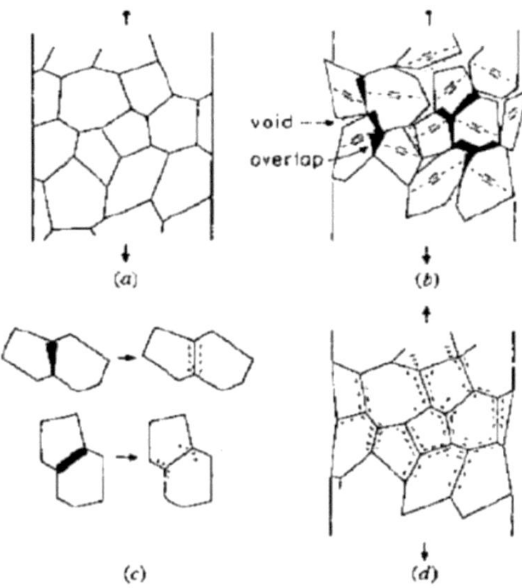

Figure 2.3: (a) Schematic diagram of the plastic deformation of a polycrystalline metal in tension, (b) voids and overlaps appear if the grains are deformed as unconstrained single crystals, (c) and (d) voids and overlaps are eliminated by geometrically necessary dislocations. (Reproduced from Ref. [20])

The simplest derivation of flow stress term containing the density of GNDs from the "classical" Taylor flow stress law can be expressed as [23]:

$$\tau = \tau_0 + \alpha\mu b\sqrt{\rho_s + \rho_g} \qquad (2.3)$$

where ρ_s is the SSDs density, ρ_g is the GND density. τ_0 is the critical resolved shear stress (the friction stress as dislocation density is zero), α is a constant related to the crystal and grain structure, μ the shear modulus and b the magnitude of the Burgers vector.

Duan et al. [24] elucidated the relationship between the total dislocations ρ_t and the statistically stored dislocation density ρ_s as well as the geometrically necessary dislocation density ρ_g. He proposed a model where the total dislocation density is expressed as:

$$\rho_t^{\mu} = \rho_s^{\mu} + \rho_g^{\mu} \qquad (2.4)$$

The parameter μ can be $\mu = 1$, $\mu = 2$, or $\mu < 1$, respectively. The μ-values indicate magnification ($\mu < 1$) or weakening ($\mu > 1$) of the significance of ρ_g when there is an increase in flow stress. However, the manipulations of μ do not have any physical justification since a plausible dislocation model is lacking.

In the last years, a number of strain gradient plasticity theories (SGP) have been developed based on the GNDs to account for the size effects. These theories in general can be divided into two classes:

- lower order strain gradient theories, which deal with conventional stresses, equilibrium equations and boundary conditions (e.g., Arsenlis and Parks [25], Busso et al. [26], Acharya and Bassani [13], Acharya and Beaudoin [27], Bassani [28], Beaudoin and Acharya [29], Evers et al. [30])

- higher order gradient theories, which have additional stress quantities and boundary conditions (e.g., de Borst and Mühlhaus [31], Fleck and Hutchinson [2, 3, 32], Fleck et al. [16], Gao et al. [8, 33], Chen and Wang [5, 6], Huang et al. [9, 34], Qiu et al. [10, 35], Gurtin [14, 36], Hwang et al. [37, 38], Mariano et al. [39], Wang et al. [40]).

Table 2.1 lists the significant strain gradient plasticity theories recently raised by different research groups. These theories are used to fit different experiments where size effects are observed.

Table 2.1: Review of recent strain gradient plasticity theories (SGP)

<table>
<tr><td rowspan="7">high order strain gradient theories</td><td colspan="1">strain gradient plasticity
theories</td><td>fitted experiments</td></tr>
<tr><td>'symmetric stress' gradient plasticity
Aifantis, E. C. (1984) [41]
Aifantis, E. C. (1987) [42]
Aifantis, E. C. (1992) [43]
Aifantis, E. C. (1999) [4]
Mühlhaus, H.B., Aifantis, E.C. (1991) [44]</td><td>micro-torsion [45] (Morrison,1939)
metal matrix composites [46] (Zhu et al., 1997)

micro-bending [47] (Richards, 1958)</td></tr>
<tr><td>CS: strain gradient couple stress plasticity
Fleck, N. A., Hutchinson, J. W. (1993) [2]
Fleck,N.A, Muller G.M., et al. (1994) [16]</td><td>micro-torsion [16] (Fleck et al., 1994)
micro-bending [48] (Stolken et al, 1998)
crack tip fields [49, 50] (Xia et al., 1996;Huang et al., 1997)</td></tr>
<tr><td>SG: Stretch and rotation gradient plasticity
Fleck, N. A., Hutchinson J. W. (1997) [3]
Fleck, N. A., Hutchinson, J. W.(2001) [32]</td><td>micro-indentation [51] (Begley et al., 1998)
grain boundary effects in bi-crystals [52] (Shu, et al., 1999)</td></tr>
<tr><td>MSG: Mechanism-based strain gradient plasticity
Nix, W. D., H. Gao (1998) [53]
Gao, H., Huang Y., et al. (1999) [8]
Huang, Y., Gao H., et al. (2000) [9]
Qiu, X., Huang Y., et al. (2003) [10]</td><td>crack tip fields [54] (Jiang et al., 2001)
boundary-layer effect on the crack tip field [55] (Shi et al., 2001)
void growth [56] (Liu et al., 2003)
micro-indentation, micro-torsion, and micro-bending [33] (Gao et al., 1999)</td></tr>
<tr><td>Gurtin's gradient theory
Gurtin, M. E. (2000) [36]
Gurtin, M. E. (2002)[14]
Gurtin, M. E., Anand L. (2005) [57]
Gurtin, M. E., Anand L. (2005) [58]</td><td></td></tr>
<tr><td>A unified treatment of strain gradient plasticity
Gudmundson P. (2003) [59]</td><td>biaxial tension of a thin film [60] (Venkatraman et al., 1992)
micro-torsion [16] (Fleck et al., 1994)
micro-indentaion [53] (Nix et al., 1998)</td></tr>
</table>

<table>
<tr><td rowspan="5">Low order strain gradient theories</td><td>Acharya and Bassani's gradient theory
Bassani, J. L. (2001) [28]
Acharya, A., Bassani J. L. (2000) [13]
Acharya, A., Bassani J. L. (1996) [61]
Beaudoin, A., Acharya A. J. (2001) [29]
Acharya, A., Beaudoin A. J. (2000) [27]</td><td>micro-torsion [16, 62] (Luo , 1998; Fleck et al., 1994)
crystal plasticity and composite hardening [63] (Cleveringa et al., 1997)
micro-indentation [64] (Ma and Clarke, 1995)</td></tr>
<tr><td>Chen's gradient theory
Chen, S. H., Wang T. C. (2000) [5]
Chen, S. H., Wang T. C. (2002) [6]
Chen, S., Wang T. (2001) [65]</td><td>micro-torsion [16] (Fleck et al., 1994)
micro-bending [48] (Stolken et al., 1998)</td></tr>
<tr><td>TNT: Taylor-based nonlocal theory of plasticity
Gao, H., Huang Y. (2001) [7]</td><td>micro-indentation [66] (McElhaney et al., 1998)
micro-torsion [16] (Fleck   et al., 1994)
micro bending [48] (Stolken et al., 1998)</td></tr>
<tr><td>CMSG: Conventional theory of Mechanism-based strain gradient plasticity
Huang, Y., Qu S., et al. (2004)[11]</td><td>micro-bending [48] (Stolken et al., 1998)
micro-torsion
void growth</td></tr>
<tr><td>DDSG: Dislocation density based strain gradient model
Brinckmann, S., Siegmund T., et al. (2006)[12]</td><td>uniaxial tension
one-dimensional strain gradient example
void growth</td></tr>
</table>

SGP theories can model size effects occurring in non-homogeneous plastic deformation involving (large) plastic strain gradients. However, size effects have also been observed in uniaxial experiments, in the absence of strain gradients, and hence another explanation must exist. Additionally, the plastic strain gradient theories are insufficient in explaining the size effect at yielding since GNDs do not exist up to the onset of plastic deformation. Moreover, all these dislocation / continuum models contain an internal material length scale. The physical meaning of this internal length scale and its relation to the measurable microstructural dimensions mostly remain unclear.

2.3 Size effects

Materials often show notable size dependencies at small sizes in the sense of: "smaller is stronger". In general, the observed size effects can be categorized as 'intrinsic' and 'extrinsic' effects. The intrinsic size effects are related to microstructural constraints, such as Hall-Petch effect and composite strengthening. The extrinsic size effects are related to plastic strain gradients, occurring in micro-bending [48], micro/nano indentation [64, 66-68], micro-torsion experiments [16]; or related to geometrical dimensions in the absence of strain gradients, for instance in thin films [69-71] or in micro-compression of pillars [72]. In some cases, the intrinsic and extrinsic size effects exhibit interactions between each other [73].

2.3.1. Intrinsic size effects

The refined grain sizes and second phase particles as composite in metal matrix are often connected with the observed size effect due to the microstructural constraints.

Hall-Petch effect

The presence of grain boundaries has an impact on the deformation behavior of a material by serving as an effective barrier to the movement of dislocations. In the early 1950s, Hall [74] and Petch [75] independently reported that the strength of iron is correlated to the grain size. Experimentally, it was found that the flow stress increased inversely with the square root of the grain size. This relationship is mathematically expressed in the form of:

$$\sigma_y = \sigma_0 + kd^{-1/2} \qquad (2.5)$$

where σ_y is the yield stress, σ_o is the overall resistance of lattice to dislocation movement (so called friction stress), k is the strengthening coefficient and d is the average grain size.

A possible derivation of Equation 2.5 originates from to the pile-up model of Eshelby et al. [76]. As shown in Figure 2.4, an array of dislocations is located between the dislocation source and the grain boundary (marked as an obstacle) with a distance L. The number of dislocations (n) in the pileup is given by:

$$n = \frac{\alpha \tau_s d}{\mu b} \tag{2.6}$$

Where α is a constant, τ_s is the average resolved shear stress in the slip plane, d is the grain size, μ is the shear modulus and b is the magnitude of the Burgers vector.

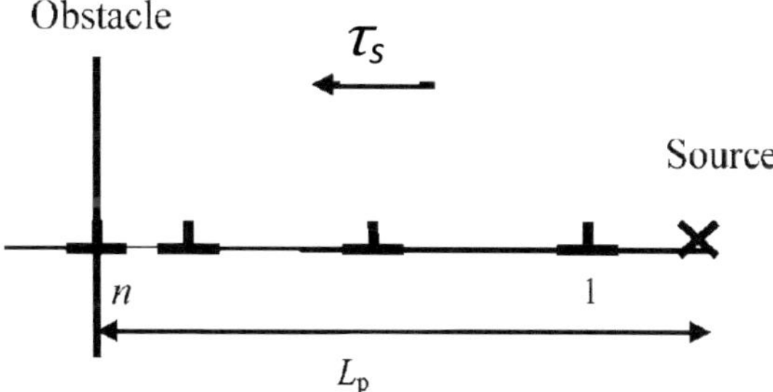

Figure 2.4: Schematic illustration of dislocation pile-up between a dislocation source and an obstacle (Reproduced from Ref. [21]).

The stress acting on the lead dislocation is found to be n times higher than τ_s. When this local stress exceeds a critical value, the blocked dislocations are able to glide past the grain boundary. However, although the pile-up model is well-defined and widely accepted, pile-ups are rarely observed, particularly in face centered cubic (fcc) metals [77]. Therefore, there is a growing consensus that the pile-up of dislocations induced by grain boundaries may not be responsible for the Hall-Petch effect.

Later, there was an important concept that grain boundaries act as dislocation sources which was introduced by Mott and Li [78, 79]. Li proposed a Hall-Petch-type relationship [80] based on grain-boundary ledges acting as dislocation sources. Figure 2.5(a) shows a schematic diagram of this model. The flow stress inside a grain can be expressed as proportional to the square root of the dislocation density, which is proportional to the number of grain boundary ledges per unit volume. However, this model is unclear how it is extrapolated to very small grain sizes since the relationship between flow stress and dislocation density has to be broken down when only a few dislocations exist per grain.

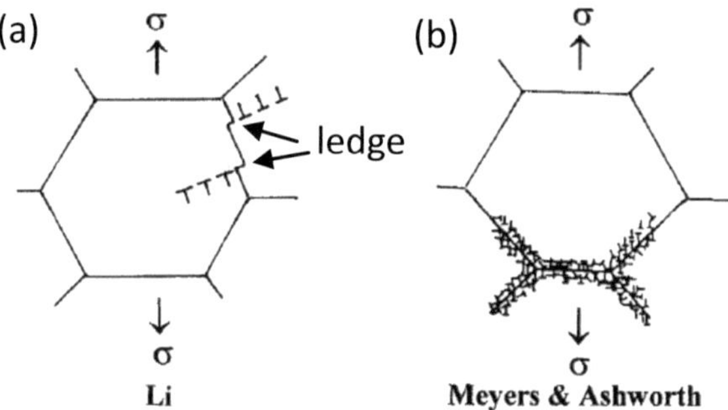

Figure 2.5: The models proposed (a) by Li, where grain-boundary ledges act as dislocation sources and (b) by Meyers and Ashworth, where generation of dislocations at grain boundaries refines the grains. These models are used to explain the Hall-Petch behavior (Reproduced from Ref. [82]).

Meyers and Ashworth [81] proposed a model based on the generation of dislocations at grain boundaries by grain refinement. In this model, as the material deforms, an elastic anisotropy results in stress concentrations in the boundary regions. Dislocations are further generated and a hardened reinforcing second-phase network is formed, as shown in Figure 2.5(b).

The dislocation density models were further developed by Conrad [83], Ashby [20] and Chia [84]. These models are based on Taylor hardening relationship between shear stress and dislocation density ρ. It was proposed that the dislocation densities (ρ) increase with decreasing grain size (d). In this case, these models give a relationship between flow stress and grain size similar to Hall-Petch theory.

The Hall-Petch relation predicts an increase in strength for smaller grain sizes when grain boundaries act as obstacles to dislocations [74, 75, 85]. However, "inverse" Hall-Petch effects appear when the grain size is smaller than 10-20 nm [86, 87], as shown in Figure 2.6. It illustrates softening takes place at grain sizes below 20 nm due to the activation of grain boundary-assisted deformation. Grain boundary-mediated deformation mechanisms prevail in this regime [88]. A crossover regime exists when the grain sizes are approximately between 20 nm and 30-100 nm. Some alternative plastic deformation mechanisms, like grain-boundary sliding, partial dislocation emission and absorption at grain boundaries [89-101] engage since the grain boundaries cannot accommodate multiple lattice dislocations in this dimension.

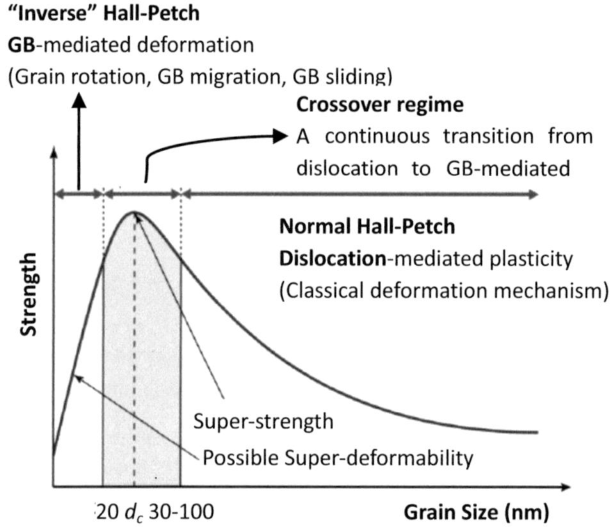

Figure 2.6: Strength of polycrystalline materials as a function of grain size: Hall-Petch relation and transition to "inverse" Hall-Petch (Reproduced from Ref. [88]).

Particle-reinforced composites

The size effect due to particle strengthening is commonly exhibited in particle reinforced metal matrix composites. The hard and brittle reinforcement particles restrict the movement of dislocations. This leads to the pile-up of the dislocations and the increase of dislocation densities [102, 103]. Hence, the stiffness and strength of the materials are enhanced. This particle size effect is closely related to the Orowan looping. When a dislocation approaches an array of hard obstacles, it is unable to cut through the particle. Instead, it loops around individual particles, as shown in Figure 2.7. The stress required for the dislocation to loop is [104]:

$$\tau = \mu b / l \qquad (2.7)$$

Where μ is the shear modulus, b is the magnitude of the Burgers vector and l is here the distance between the particles.

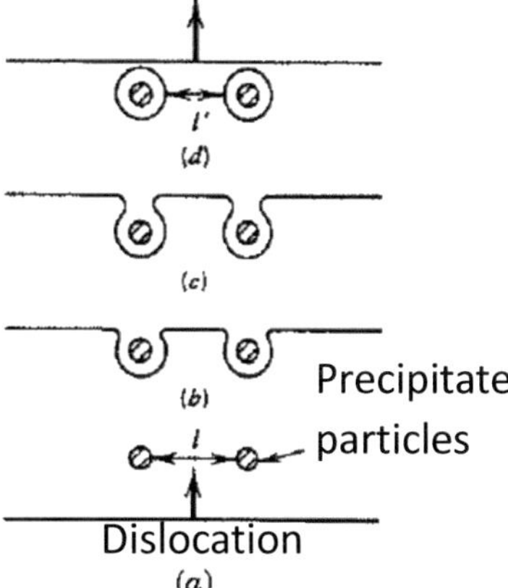

Figure 2.7: Schematic illustration of the reduced particle spacing during looping (Reproduced from Ref. [105]).

Experimental results revealed that both volume fraction of particles and particle size have an influence on the size effect [106-108]. For a given particle size, the effective distance l between two adjacent precipitates decreases with increasing the number of dislocation loops surrounding the particles, as shown in Figure 2.7. As such, strain hardening is induced by the dislocation looping mechanism [109]. For a given volume fraction of second-phase particles, l increases as the precipitates grow larger. Consequently, the stress necessary for dislocations to loop around precipitates decreases with increasing particle size.

2.3.2. Extrinsic size effects

Extrinsic size effects arise from either gradients in plastic strain or geometrical dimensions in the absence of strain gradients, such as from small sample size or small strained volume during the deformation process.

Size effects related to gradients in plastic strain

The extrinsic size effects related to gradients in plastic strain were observed in micro-torsion (refer to Section 2.1), micro-bending, mirco/nano-indentation experiments etc..

A classical micro-bending experiment was performed by Stölken and Evans on micro nickel foils with thicknesses between 12.5 and 100 μm [48]. They observed a significant increase in the bending strength with decreasing foil thickness, as shown in Figure 2.8(a). The results are well described by the SGP theory [110]. GNDs are generated to accommodate lattice curvature, which obstruct the motion of SSDs and hence, enhance work hardening in the material. Moreover, the results can be also explained by the critical thickness theory (CCT), which was introduced by Frank & van der Merwe in 1949 [111]. Up to a critical thickness, misfit dislocations are generated at the thicker foil to reduce the strain by plastic relaxation and this leads to a low flow stress [112]. However, the grain size of the different foils was not taken into account in the experiment from Stölken and Evans. Meanwhile, there is a lack of data around the elastic-plastic transition region. Moreau et al. [113] used the same micro-bending technique as Stölken and Evans. They observed an increase in the yield strength with decreasing foil thickness. Besides, they obtained accurate results in the low strain region and around the yield point for nickel foils with same grain sizes. Motz and co-workers [114] reported on micro-bending experiments on copper beams, machined by focused ion beam technique, with equal thickness and width in the range 1-7 µm. The strength of beams was found to vary inversely with beam thickness. A new mechanism has been proposed that a dislocation pile-up appears in the center of the beam. This leads to additional stresses that enhance the flow stress value, as shown in Figure 2.8(b). In other words, the dislocation pile-up in the center has a greater impact in thinner beams, which results in higher strength.

The indentation size effect (ISE) is well-known as shown in many experiments that the hardness decreases as the depth of indentation increases. It is more significant at depths of less than approximately 1 µm. In metals, the factor of the decrease in hardness with depth of indentation is typically two or three [64, 115]. ISE was observed for both using 'sharp' (e.g., knife edge, conical or pyramid) and spherical indenters. ISE was first reported in gold single crystals in the early 1970's that hardness could be enhanced by decreasing the contact diameter from 0.5 mm to 0.2 µm with a Vickers indenter [116]. Later, Stelmashenko et al. [67] showed similar hardness increase at shallow depths in a single crystal tungsten with various orientations, as illustrated in Figure 2.9(a). The hardness, H, was given by:

$$H = A\alpha\mu b(\rho_0 + \frac{\cot\beta}{bd})^{1/2} \qquad (2.8)$$

where A and α are non-dimensional coefficients of a constraint and Taylor hardening, μ is the shear modulus, b is the magnitude of the Burgers vector, ρ_0 is the background dislocation density, $\cot\beta$ is the wedge shape, and d is the diagonal of an indent with a Vickers's hardness diamond indenter.

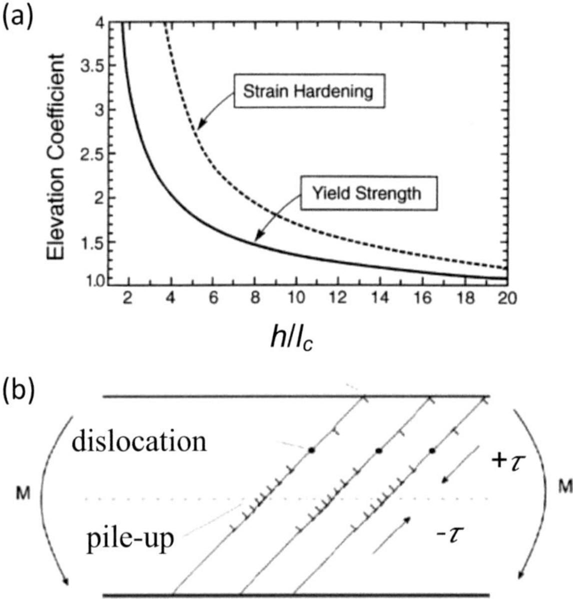

Figure 2.8: (a) The impact of foil thickness on yield strength and strain hardening (h is the foil thickness, lc is length scale) [48], (b) schematic illustration of dislocation pile-up mechanism in the deformation of micro-beams (Reproduced from Ref. [114]).

In contrast to Vicker's pyramid indenters, spherical indenters have the advantage to create a larger contact area as well as a more homogeneous stress field at smaller penetration depth. The diameter of the sphere is the most important length scale for spherical indenters.

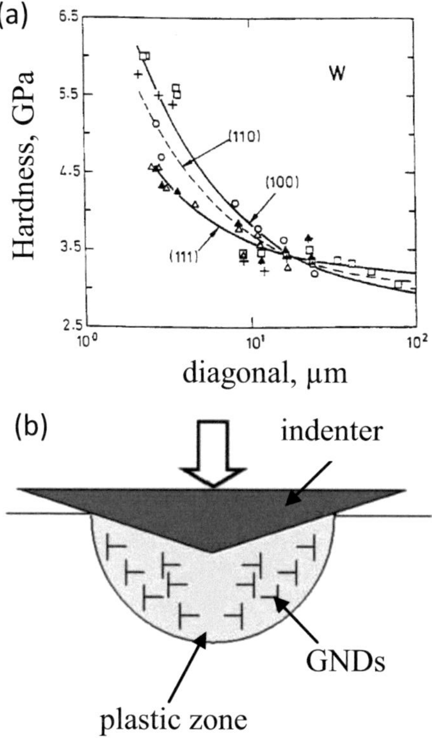

Figure 2.9: (a) Hardness data for tungsten single crystals at three orientations relative to the direction of the Vickers indents [67], (b) geometrically necessary dislocations underneath an indenter (Reproduced from Ref. [110]).

Lim and Chaudhri [117] first demonstrated the ISE on annealed copper samples using spherical indenters, where higher contact pressures occurred when smaller radius indenters were used. Similar results were obtained for iridium by Swadener et al. [118]. The ISE for crystalline materials can be explained in terms of the local dislocation hardening based on geometrically necessary dislocations. Figure 2.9(b) illustrates a schematic depiction of dislocations underneath an indenter. GNDs are required to accommodate the lattice distortion when the material underneath the indenter is pushed into the substrate during deformation. As the depth or the diameter of indentation decreases, the GND density increases and results in higher hardness.

In summary, the strain gradient theories can successfully describe the size effects observed in micro-torsion, micro-bending and micro/nano-indentation investigations, although different strain gradient models are required for various experimental observations. In some cases, other theories, such as critical thickness theory, are also used for describing the observed size effects.

Size effects related to geometrical dimensions in absence of strain gradients

Size effects were also detected in the absence of strain gradients related to geometrical dimensions, such as in thin films [119] or micro-compression and tension of micron/submicron pillars [72].

A size-dependency was detected in compression or tension of single crystalline micron- and submicron-pillars. The yield strength of sub-micron single crystal metallic pillars and wires increases with decreasing pillar diameter. Figure 2.10(a) shows a compression test on a micro-pillar using a flat indenter tip. An universal scaling law was proposed for the strength of fcc metals by Dou and Derby [120]. A power-law dependence was found between the flow stress and sample size, as expressed:

$$\frac{\sigma_{crss}}{\mu} = A(\frac{d}{b})^m \tag{2.9}$$

where μ is the shear modulus, σ_{crss} is the resolved shear stress onto the $\{111\}<110>$ slip system, d is the pillar diameter, b is the magnitude of the Burgers vector, m is power-law slope and A is a constant. For fcc metals, $A = 0.71$ and $m = -0.66$.

Figure 2.10(b) illustrates the correlation between strength and pillar diameter. The resolved shear stress σ_{crss} normalized by shear modulus μ (resolved onto the $\{111\}<110>$ slip system) is plotted against pillar diameter d normalized by the Burgers vector b of the dominant slip system. The unified power-law slope ($m = -0.66$) can be detected for fcc metals (Au, Al, and Ni). In this study, bcc metals such as Mo and Mo alloys, were also found to obey the fcc correlation.

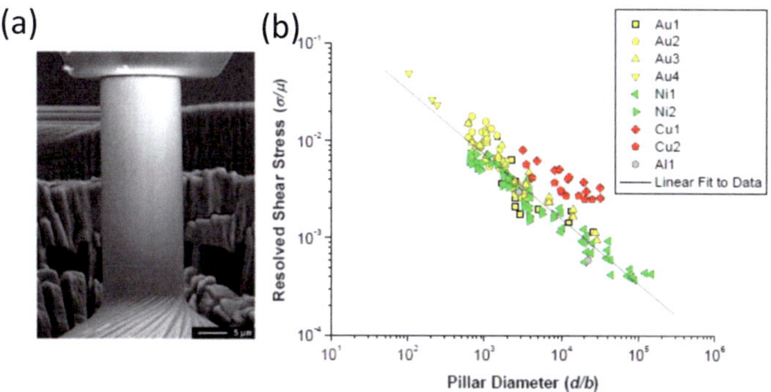

Figure 2.10: (a) Micro-pillar under compression test using a flat indenter tip [126], (b) resolved shear stress normalized by shear modulus (resolved onto the $\{111\}<110>$ slip system) for fcc micro-pillars, plotted against pillar diameter normalized by Burgers vector (Reproduced from Ref. [120]).

It is hypothesized that the deformation mechanisms for the size dependency in the micro-pillars are controlled by the heterogeneous nucleation of dislocations from surface imperfections (surface nucleation theory [121, 122]) or sparsely distributed sources (single-arm source theory [123-125]). However, there is yet no unified plasticity theory established.

Size effects were also observed in thin films on bulk substrates as well as multilayered coatings for thin films. The flow stress σ_y is found to increase with decreasing film thickness h_f [69-71], as shown in Figure 2.11(a). Based on the critical thickness theory [127, 128], one can describe the strength of a thin film as a function of its thickness:

$$\sigma_y \propto \frac{1}{h_f}\ \ln(\frac{\alpha h_f}{b}) \qquad (2.10)$$

where σ_y is the yield stress, h_f is the film thickness, b is the magnitude of the Burgers vector, and α is a constant.

Figure 2.11(b) shows the mechanism of dislocation motion in a single metal film subjected to a simple in-plane biaxial loading. Since the dislocation is confined to move in the film but not in the substrate, a 'misfit' dislocation will be deposited near the film/substrate interface as the dislocation glides on its slip plane. Matthews et al. [19] first reported that a critical stress is needed to create misfit dislocations by the process. The strengthening arises either from the dislocation channeling (resistance to motion of a dislocation by forming compact misfits as illustrated in Figure 2.11(b)) [119] or from the dislocation blocking (resistance to motion of a dislocation such as by a misfit lying on an intersecting slip plane) [129]. By increasing sample size, the governing mechanism changes from nucleation of individual partial dislocations to nucleation of individual full dislocations, interaction and multiplication of these dislocations [130]. Kraft et al. [130] confirmed that there is no single mechanism with a universal power-law exponent for the observed size effect in thin films.

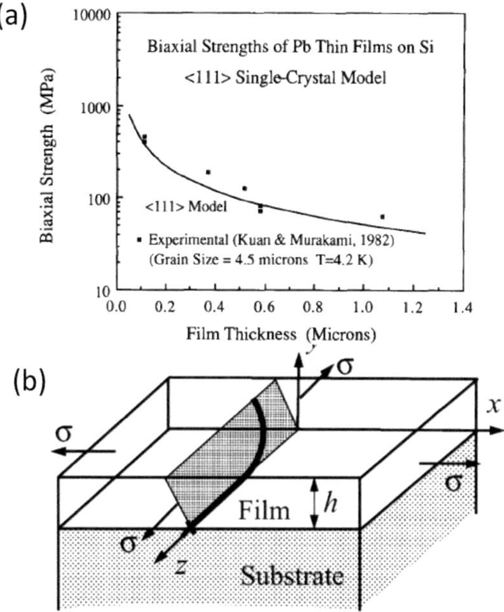

Figure 2.11: (a) Biaxial strengths of Pb thin films on Si substrates as a function of film thickness [119], (b) mechanism of dislocation motion in a thin single crystal film on a substrate. A 'misfit' dislocation is deposited at the film/ substrate interface as the dislocation glides in its slip plane (Reproduced from Ref. [131]).

So far, the mechanisms such as dislocation nucleation, dislocations starvation and others prevail for the size effect related to geometrical dimensions in absence of strain gradients. However, there is no unified theory established and a number of different underlying mechanisms are responsible for the observed size effect.

2.3.3. Interaction between intrinsic and extrinsic size effects

Intrinsic and extrinsic size effects are not independent. The microstructural size effect (intrinsic) is often observed to superimpose on the geometrical size effects (extrinsic).

In 1961, Armstrong identified the coexistence of a structure size effect and a grain size effect [132]. However, the Hall-Petch effect was not systematically studied for small structures. Venkateswaran and Bravman [133] investigated aluminum films on silicon substrates. Similar dependence of flow stress (σ_y) on both the film thickness (h_f) and grain size (d_g) was observed. However, there were only two grain sizes (0.9 µm and 1 µm) investigated in this work, which was not enough to distinguish the grain size dependency in the form of $d^{1/2}$ or d^{1}. Ehrler et al. [73] studied the effect of grain size on the bending behavior of thin metal foils with a thickness (h_f) of 10, 50 and 125 µm, respectively, and grain sizes (d_g) ranging from 6 to 200 µm. Two different stress dependencies were found. At low strain levels:

$$\sigma \propto \sqrt{\frac{1}{h_f} + \frac{1}{d_g}}$$ (2.11)

and at high strain levels:

$$\sigma \propto \frac{1}{h_f} + \frac{1}{d_g}$$ (2.12)

Equations (2.11) and (2.12) suggest that the size effect is driven by finite strained volume, which may be delimited by grain boundaries as described in the classic Hall-Petch effect, or by free surfaces as observed in the gold pillar experiments, or even by a strain gradient.

Hou et al. [134] observed an interaction between grain size effect and indentation size effect. Experiments with spherical indenters of different radii were carried out on both single and polycrystalline copper. The grain size varied from 1.15 μm to '∞' (single crystal) and the indenter radii varied from 0.82 to 50 μm. The dependency of the stress (σ) from contact radius (a) and grain size (d_g) is expressed as:

$$\sigma \propto \sqrt{\frac{1}{a} + \frac{1}{d_g}} \qquad (2.13)$$

In these investigations, interactions between Hall-Petch effect and geometrical size effects were observed. However, the relationship between the effects has not been quantified.

It was observed that in metallic wires with diameters ranging from 10 to 100 μm, grain size hardening becomes less effective for decreasing wire diameters when the grain size is a reasonable fraction of the wire diameter [135, 136]. Chen [135] reported that the strength of the micro wires was likely to be affected by three factors: (i) grain size d (Hall – Petch); (ii) specimen diameter t; and (iii) shape as characterized by the ratio t/d, which can be expressed as:

$$\sigma = \sigma_1(t) + \sigma_2(t/d) + kd^{-1/2} \qquad (2.14)$$

However, in most studies the interplay of these different size effects is not fully taken into account. Many theories exist, each addressing a small domain of the experimentally observed size effects and formulating its respective mechanisms. However, most theoretical explanations do not consider the interaction between different effects. The same problem exists in strain-gradient plasticity theories. Besides, the individual impact of the effects on the mechanical behavior of micro structures under different loading conditions and at different strain levels, requires further investigation. This will offer a better insight of the underlying mechanisms in size effects.

2.4 Superimposing effects

In order to determinate the pure size effects, superimposing effects should be separated. The most important superimposing effects are caused by: (i) texture differences, (ii) different dislocation densities, and (iii) strain rate sensitivity. Hence, comparable microstructures and test conditions are required.

2.4.1 Texture

Texture is the distribution of crystallographic orientations of a polycrystalline material. The materials with fully random orientations is said to have isotropic properties. If the crystallographic orientations have some preferred orientation, the material has a texture and shows anisotropic behavior. The degree of texture (weak, moderate or strong) is dependent on the percentage of crystals having a preferred orientation. Texture is caused by materials processing, forming and heat treatments. For cold drawn wires, a texture along the cold drawing direction is often detected.

Young's modulus (E) is indicative of possible textures in materials due to elastic anisotropy. It is a measure of the stiffness and describes the tendency of an object to deform along an axis when opposing forces are applied along that axis. Young's modulus is defined as the ratio of the uniaxial stress over the uniaxial strain, which can be experimentally determined from the slope of the elastic regime of a stress-strain curve obtained from a tensile test.

In anisotropic materials, Young's modulus may have different values depending on the direction of the applied force with respect to the material's structure. The Young's modulus of gold for example is varying between $E_{<111>} = 117$ GPa and $E_{<100>} = 43$ GPa [22]. For a given lattice, the yield strength and hardening will be also affected by the orientation of an applied load.

2.4.2   Dislocation density

Dislocations have a strong impact on the mechanical behavior of a material. The structures, crystallography of dislocations, as well as the energy and stress fields affect the plastic flow. In general, a cold worked metal with high dislocation density has high yield strength and low elongation at fracture.

Dislocation density can be affected by materials processing, e.g., by quenching or cold work hardening by forging, rolling and cold drawing. An effective way to change the dislocation density/structure is to heat treat materials. The occurring recovery and recrystallization process can minimize the inner energy of materials, which is related to the high dislocation densities.

In the post cold deformation state, some of the energy involved in the deformation is stored in the form of dislocations and point defects. This is known as stored energy of cold work. Any heat treatment intended to reduce or eliminate pre-deformation is termed as annealing. During annealing, the following processes are involved with increasing temperatures [137]:

1. a large reduction in the number of point defects

2. dislocations of opposite sign attract and annihilate each other

3. dislocations rearrange themselves towards lower energy configurations

4. both point defects and dislocations are absorbed by grain boundaries migrating through the material

5. a reduction in the number of grain boundaries

Figure 2.12(a) shows the microhardness plotted against annealing temperature on gold alloys. The annealing processes have three stages: recovery (Processes 1-3), recrystallization (Process 4, possibly including Process 3 whereby recrystallization nuclei form) and grain growth (Process 5).

Recovery includes all processes which do not require the movement of a high angle grain boundary to release stored energy [138]. Typically, recovery processes involve the rearrangement of dislocations in order to lower the inner energy of a material, for example by forming low-angle sub-grain boundaries [139].

Recrystallization is the formation of a new grain structure in a deformed material by formation and migration of high angle grain boundaries, driven by the stored energy of deformation. High angle grain boundaries separate grains with orientation difference greater than 10-15° [139]. The recrystallization temperature is directly related to the degree of cold-work hardening. The higher the degree of cold-work hardening, the more deformation energy is stored in the lattice. Consequently, a lower energy, i.e., a lower annealing temperature is required to activate recrystallization, as shown in Figure 2.12(b). Cold work hardened materials exhibit in general the problem, that the degree of pre-deformation decreases from the surface towards the center. Consequently, the degree of recovery/recrystallization for the materials from the surface to center will be different.

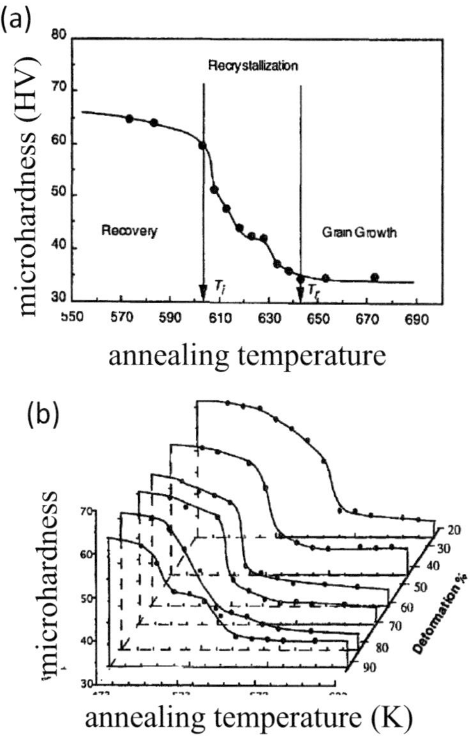

Figure 2.12: (a) Microhardness (HV) of gold alloys annealed for 1 hour in dependence of the temperature (recrystallization starting at T_i and finishing at T_f), (b) Recrystallization map showing the correlation among microhardness (HV), temperature and degree of pre-deformation (Reproduced from Ref. [140]).

Grain growth is a process involving the migration of grain boundaries, where the driving force for migration is the reduction of the grain boundary area itself. The growth of the new grains depends on both the mean stored energy [141] and the frequency of new grains re-acquiring a low mobility boundary by meeting similar orientations in the deformed state [142, 143].

2.4.3 Strain rate dependency of plasticity

For many materials, it is known that the flow stress and/or the hardening are increasing with increasing strain rate $d\varepsilon/dt$, as shown in Figure 2.13. The sensitivity depends on the structure of a material. Body-centered-cubic (bcc) metals (e.g., iron, chromium, molybdenum and tungsten) are much more sensitive to $d\varepsilon/dt$ compared to fcc metals (e.g., aluminum, copper, gold and nickel). Polymeric solids reveal also strong strain rate sensitivity.

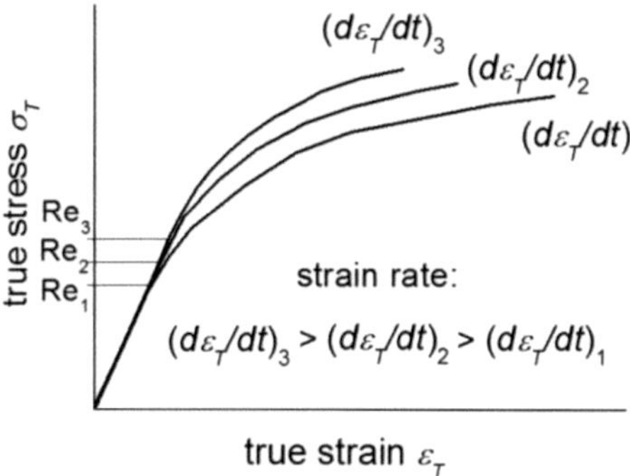

Figure 2.13: Schematic diagram of the strain-rate behavior in dependence of the strain rate $d\varepsilon_T/dt$.

The relationship between true stress and strain rate is given by [105]:

$$\sigma_T = K(\frac{d\varepsilon_T}{dt})^m \tag{2.15}$$

where σ_T is the true stress, K is a material constant, $d\varepsilon_T/dt$ is the true strain rate and m is the strain-rate sensitivity factor. For most metals, m is low and varies between 0.02 and 0.1. Under certain conditions wherein $m > 0.3$, a given material may exhibit a significant degree of strain-rate sensitivity in combination with super-plastic deformation behavior. At the limit, where $m = 1$, the stress-strain-rate material response is analogous to Newtonian viscous flow.

Depending on the nature of the test or service condition, strain rates may vary by more than a dozen orders of magnitude. The common ranges for strain rate testing are between 10^{-1} and 10^{-3} s^{-1} (quasi-static) and 10 to 10^3 s^{-1} (dynamic) [144]. At low strain rates, i.e., below about 10^{-3} s^{-1}, materials behavior is characterized by its creep and stress rupture response. At strain rates between 10^3 and 10^5 s^{-1}, under impact conditions, materials may fail with a reduced fracture energy [105].

Moreover, strain rate sensitivity is also known to have a strong relationship with the grain size. Previous studies showed coarse-grained gold films [145] with thicknesses ranging from 0.8 to 2.0 μm and grain size lager than 1.5 μm are strain-rate insensitive, whereas fine-grained gold films [146] with thicknesses ranging from 0.2 to 2.0 μm and grain size around 100-500 nm have shown a strain-rate sensitive behavior. A similar strain rate sensitivity was observed in fine grained gold wires with a diameter of 25 μm and grain size around 200-500 nm [146]. Besides, the dependency of strain rate sensitivity on the grain size was also determined in investigations on microcrystalline pure Ni [147].

Figure 2.15 shows that typically the elongation at fracture decreases with increasing strain rate. However, anomalous effects of strain rates on the tensile elongation were also observed in several investigations [148-150]. Cao et al. [149] investigated coarse-grained commercially pure iron containing grain boundary micro-voids at strain rates ranging from 10^{-5} s^{-1} to 10^{-2} s^{-1} under uniaxial tension. It is found that the elongation increases abnormally with increasing strain rate. They proposed that the extending deformation of micro-voids towards the tensile stress direction contributed more to the total elongation as the strain rate increases. Xu et al. [150] performed tensile tests on a fully recrystallized 50Mo-50Re alloy at strain rates ranging from 10^{-6} to $1\ \mathrm{s}^{-1}$ at room temperature. They found that the total elongation of the alloy increased significantly with increasing strain rate. The decrease in ductility at low strain rates in this alloy is assumed to be related to the interaction between dislocations and interstitial solved atoms in the lattice. An anomalous increase of the tensile elongation was also observed to take place on Co_3Ti alloys with small grains at very low strain rate region (mostly below $10^{-5}\ \mathrm{s}^{-1}$) and also at high testing temperature (i.e., 353K) [148]. This was suggested to be associated with the oxidation reaction during the tests due to hydrogen decomposition from moisture in the air at high temperature. The oxide production on the alloy surface counteracts to the embrittlement and is responsible for the anomalous increase of the elongation. In general, grain boundary sliding is considered as a major reason for the anomalous strain-rate effect in super-plastic materials with fine grains [151-154]. Anomalous effects of strain rates on the tensile elongation were found at low strain rate region under tensile tests, however, the mechanism may vary significantly depending on the material and the testing environment.

3.   Experimental details

Section 3.1 introduces the material used in this work. Section 3.2 describes heat treatments conducted to create different microstructure. Section 3.3 outlines the methods used for microstructural characterization, such as focused ion beam microscopy (FIB) and electron backscatter diffraction (EBSD). In Section 3.4, the test set-up for tension and torsion tests and the experimental procedures for mechanical characterization will be introduced.

3.1   Material

In this study, polycrystalline gold (Au) wires with diameters of 12.5 μm (Goodfellow, Germany) and 15, 17.5, 25, 40, 60 (Haeraeus, Germany) from high purity gold (99.99 at.%) were investigated. Besides, polycrystalline aluminum (Al) wires (containing 1 wt. % Si, Heraeus, Germany) with diameters of 15, 17.5, 25 and 40 μm were also investigated. The real diameters of the wires were determined by images obtained from a scanning electron microscope (SEM). Table 3.1 lists both the nominal and measured diameters of Au and Al wires in the as-received state. Each wire was measured at more than ten different locations with an interval larger than 0.2 mm along the entire sample in order to determine the average wire diameter.

Table 3.1: Comparison of nominal and measured diameters of Au and Al wires in the as-received state

	Nominal diameter (μm)	measured diameter (μm)
Au wires	12.5	11.68 ± 0.35
	15	14.78 ± 0.02
	17.5	16.73 ± 0.17
	25	24.46 ± 0.09
	40	37.38 ± 0.15
	60	59.53 ± 0.34
Al wires	15	14.72 ± 0.08
	17.5	17.49 ± 0.03
	25	25.59 ± 0.16
	40	38.11 ± 0.10

3.2 Heat treatment

The samples for heat treatments were prepared on a frame which is made of two steel rods and high temperature cement (OMEGABOND 300 Air Set Cement). Figure 3.1 is an illustration of a sample on the heat treatment frame. The wires were fixed loosely on the frame to avoid pre-deformation caused by a thermal swelling of the frame, which could potentially occur during heat treatments.

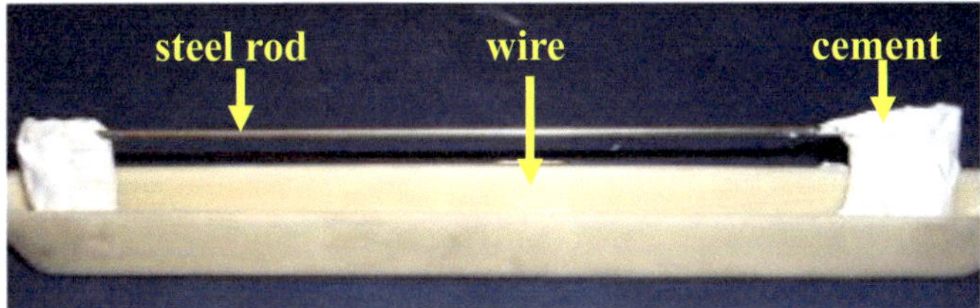

Figure 3.1: Heat treatment frame for the as-received gold wires.

Samples were annealed using a high vacuum glass-tube furnace (Heraeus, type 7/75) under a vacuum of 10^{-6} mbar. For each material and wire diameter, temperature and dwell time were systematically varied from 200°C to 560°C and 2 hours to 10 hours to change the microstructure and the resultant strength and toughness.

3.3 Microstructural characterization methods

The methods used for microstructural characterizations are mainly focused ion beam (FIB), electron backscatter diffraction (EBSD) and Auger electron spectroscopy (AES). The thickness of the oxide layer on the aluminum wires was determined from Auger Nanoprobe measurements (AES, type PHI 680) with a 10kV acceleration voltage. The application of FIB and EBSD will be introduced in Sections 3.3.1 and 3.3.2.

3.3.1 Focused ion beam microscopy

The microstructure was characterized using a Dual Beam Workstation (FEI Nova NanoLab 200) which is equipped with an electron beam column (SEM) and a focused (Ga+) ion beam (FIB) column. The channeling of the incident ions between lattice planes of a specimen results in grain orientation contrast [155]. This means that different crystal orientations have an influence in the number of secondary electrons escaping from the specimen. This enables different grains to have different contrast. Based on a distinctive channeling contrast, the ion beam can be used to visualize the grain structure of a material. Figure 3.2(a) shows the microstructure of the cylindrical surface of a gold wire, whereas Figures 3.2(b) and (c) illustrate the cross-sectional microstructures along and perpendicular to the longitudinal axis of wires, respectively.

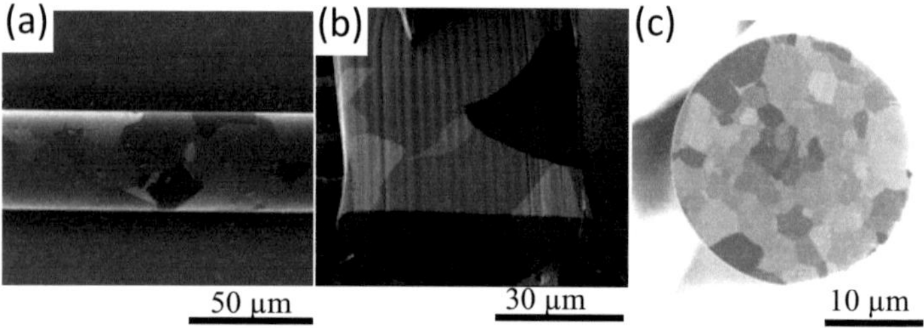

Figure 3.2: FIB images of gold wires (a) at the cylindrical surface, at a cross-section (b) along and (c) perpendicular to the longitudinal axis of a wire.

The line intercept method [156] was used to determine the grain size from the FIB images. The procedure involves drawing lines of length L randomly within a micrograph and counting the number of grains, N, which are cut by the lines, as seen in Figure 3.3(a). Whole grains are counted as 1 while grains at the end of the lines are counted as 1/2. The average grain size, d_g, can be deduced by:

$$d_g = \Sigma \left(\frac{L}{N} \right)_i / i \qquad (3.1)$$

In this study, the lines across the center of a wire were chosen to determine the average grain size since the microstructure varies from the surface to the center after annealing. This is possibly due to the different degrees of recovery/recrystallization, as shown in Figure 3.3(b). Several cross-sections of one sample cut along the length were analyzed in order to determine an accurate average grain size.

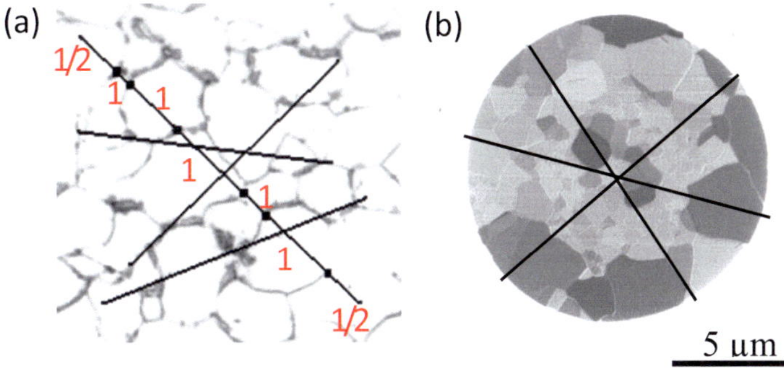

Figure 3.3: (a) Schematic illustrations of Heyn-Hilliard-Abrams intercept method for average grain size determination, (b) the modified intercept method for this study, using lines across the center of a wire.

3.3.2 Electron backscatter diffraction

Electron backscatter diffraction (EBSD) is an effective approach to investigate textures and grain orientations of polycrystalline materials. Besides, local misorientations obtained from EBSD measurements can also be used to calculate GND densities and distributions. An approach introduced by Calcagnotto [157] for GND densities calculation is used here. This approach is based on the work of Kubin and Mortensen [158]. Kubin and Mortensen [158] proposed a simple torsion model of a cylinder with a series of twist subgrain boundaries. Each sub-boundary contains two perpendicular arrays of screw dislocations oriented normal to the axis. The dislocations required to accommodate a twist angle over a unit length is found to be:

$$\rho_{GND} = \frac{2\vartheta}{ub} \qquad\qquad (3.2)$$

where ϑ is the misorientation angle, u is the unit length and b is the magnitude of the Burgers vector. The misorientation angle ϑ is related to the kernel average misorientation (KAM, local lattice curvature). The KAM was obtained by calculating the misorientation gradient of each EBSD data point relative to its first, second and third nearest-neighbor points [159]. Figure 3.4 shows the principle of GND density calculation based on the misorientation.

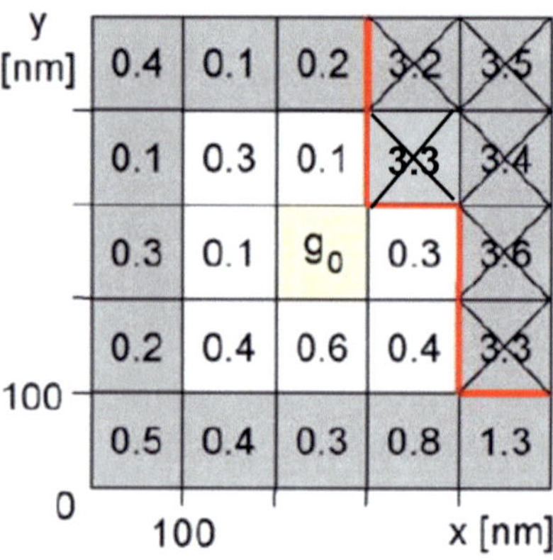

Figure 3.4: Principle of GND density calculations based on the misorientation [157]: The closest (white) and second-closest (grey) points around a given measurement point g_0 are chosen for the measurement. Misorientations that exceed the set threshold value are excluded from the calculation, as shown by the points with black crosses. Red lines signify grain boundaries.

The KAM quantifies an average misorientation based on a set of points which are closest or the closest plus the second-closest to a given measurement point g_0. A step size of 0.1 μm was used in this study to define a point when measuring the misorientation. The red lines indicate the grain boundaries. A predefined threshold misorientation value is set to be 2° in this study. When misorientations exceed this threshold value, the points are excluded from the calculation for the average misorientation, as indicated as black crosses in Figure 3.4. These points are assumed to belong to the adjacent grains or the subgrains.

3.4 Mechanical characterization

3.4.1 Test set-up

3.4.1.1 Measuring principle to determine torsion moments

Figure 3.5 schematically shows the measuring principle to determine torque curves in our test set-up. A specimen is clamped between a load cell and a rotation table under a low axial preload. At a defined height, which depends on the required initial gauge length of the specimen, a stiff cross-beam is glued to the wire. Two load sensors (atomic force microscopic (AFM)-tips, Kleindiek Nanotechnik) are positioned opposing to each other with equal distances to the specimen such that a symmetrical configuration is achieved. Since the movement of the cross-beam is inhibited, the force is transferred to the load cells and recorded by the measurement system. As the inner moment of the wire is equal to the outer moment of the cross beam, the torsion moment M_t is described as:

$$M_t = M_{d\,left} + M_{d\,right} = a \cdot F_{left} + a \cdot F_{right} \qquad (3.3)$$

where a is the distance between the center of the wire and the load sensors, F_{left} and F_{right} are the measured forces, and $M_{d\,left}$ and $M_{d\,right}$ are the respective torque moments. The subscripts 'left' and 'right' indicate the left and right load sensor, respectively. A more detailed description and the general capabilities of the test set-up are presented elsewhere [160, 161].

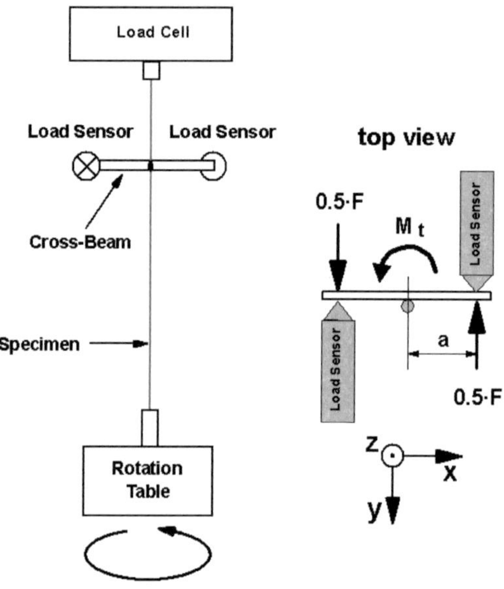

Figure 3.5: Schematic illustration of the measuring principle to determine the torque during deformation in torsion.

3.4.1.2 Test set-up

Figure 3.6 shows the in-house test set-up which is based on a small table-top tensile testing machine (Zwick, type Z2.5) with a 10 N load cell [160]. Both tension and torsion tests were performed at room temperature using this test set-up. As we can see from Figure 3.6(d), a rotation table (PI -Physik Instrumente, type M-038.PD1, with a resolution of 27 μrad and without angle restriction) is mounted between two x-y tables on a base plate. The large x-y table has a resolution of 1 μm (OWIS, type KT 90-D56-EP) and is positioned under the rotation table on the base plate. The small x-y table is made from two single linear tables with a resolution of 200 nm (mechOnics, type MS 30) and is positioned on the hollow copper block that is above the rotation table, as shown in Figure 3.6(c). The specimen is clamped between two grips, as shown in Figure 3.6(b). One of the grip is installed under the load cell, as seen from Figure 3.6(a). The other one is installed above the small x-y table. Two types of load sensors (atomic force microscopic (AFM)-tips, Kleindiek Nanotechnik) are used to determine torsion moments:

- type FMT 400 (Fmax ~ 100 μN) for thinner wires and

- type FMT 120 (Fmax ~ 500 μN) for thicker wires

The AFM-tips are sensitive to determine torsion moments even in the nNm regime. A system of x-z stages with a resolution of 50 nm (PI, type M-112.1DG), as shown in Figure 3.6(d). Linear y-positioners (nano motors, Kleindiek Nanotechnik) with a resolution less than 1 nm are installed at the top of the x-z stages to adjust the load sensors back and forward. The load sensors are positioned at the tip of the y-positioners, as shown in Figure 3.6(b).

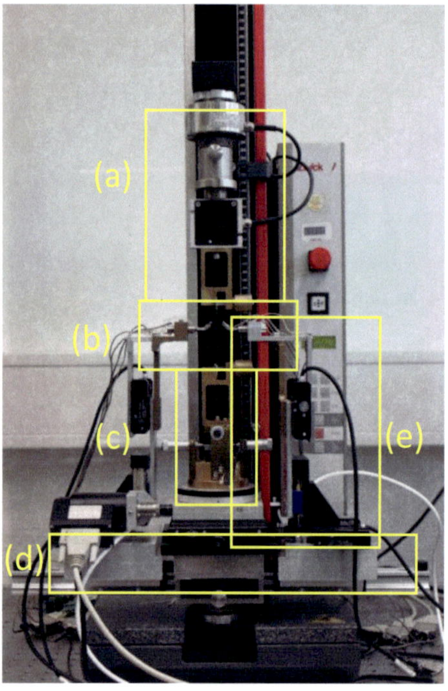

Figure 3.6: Overview of the test set-up for both tensile and torsion tests. (a) Upper part of the test set-up containing the load cell and upper specimens grip, (b) specimen and load sensors, (c) middle part of the test set-up containing lower specimen grip, helical rods, small x-y table, copper block, rotation table

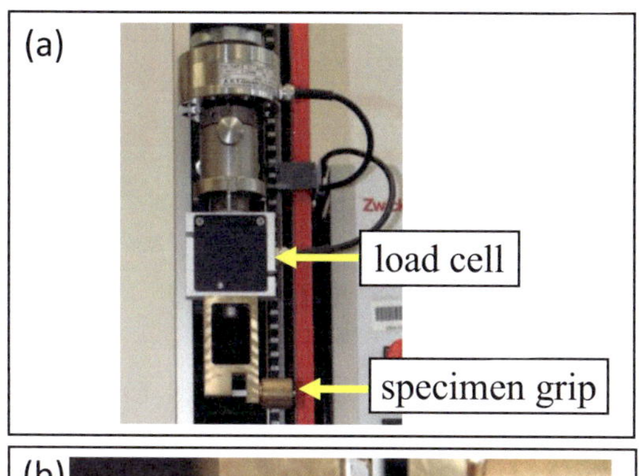

(a)
load cell
specimen grip

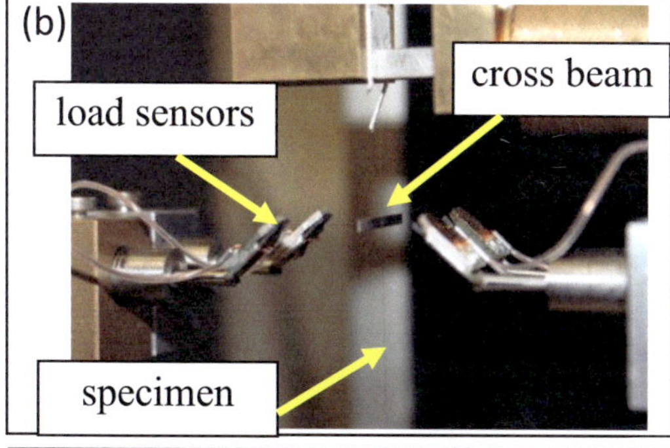

(b)
load sensors
cross beam
specimen

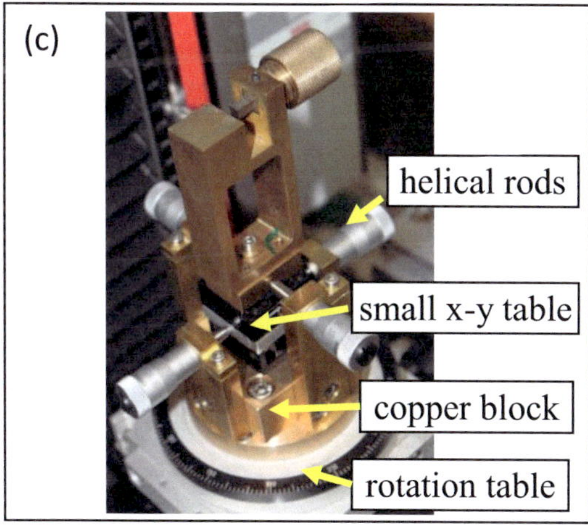

(c)
helical rods
small x-y table
copper block
rotation table

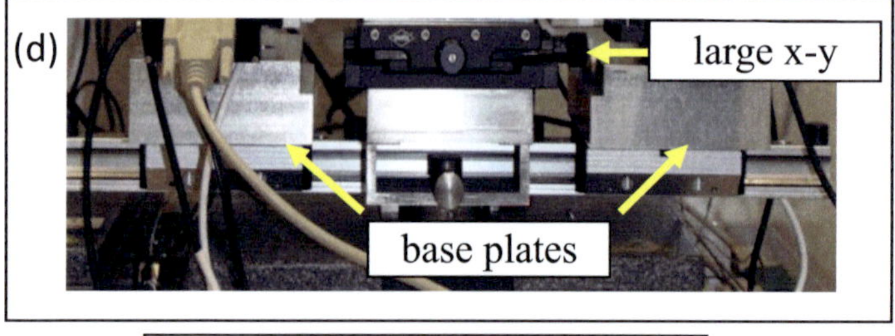

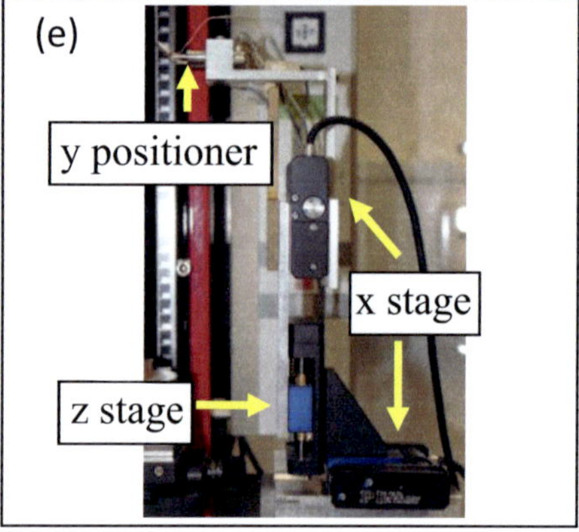

Figure 3.6: Overview of the test set-up for both tensile and torsion tests.
(d) lower part of the test set-up containing large x-y table and base plates, and
(e) x-z stages with y-positioners for the load sensors.

The set-up allows both monotonic and cyclic torsion tests, as well as combined tensile-torsion tests to be conducted at the same time. Figure 3.7(a) shows the torque curves from monotonic torsion tests. The results reveal negligible discrepancies when different test configurations (i.e., one or two tips) were used. Figure 3.7(b) shows a torque hysteresis determined from a cycling test. This reveals that this test set-up has the potential to perform cyclic tests, where mechanical properties such as Bauschinger effect can be determined.

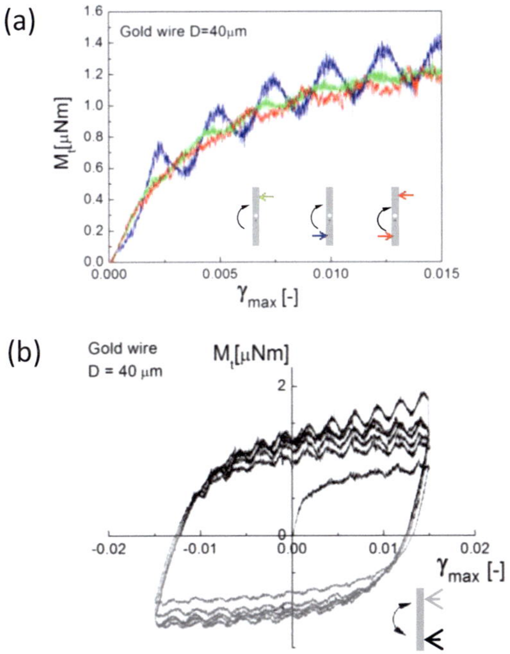

Figure 3.7: Torque curves determined from torsion tests under (a) monotonic and (b) cyclic loadings.

3.4.1.3 Optimization of the test set-up

Since the original test-setup had issues on alignment of specimens and structural stability, individual components of the set-up were modified at the beginning of this study. Figure 3.8 shows a comparison of the middle part of test set-up before and after modification. A large x-y table was added onto the integrated base plates. Four helical rods were used to fix the small x-y tables. The components between the rotation table and small x-y tables for alignment were improved, which will be detailed below.

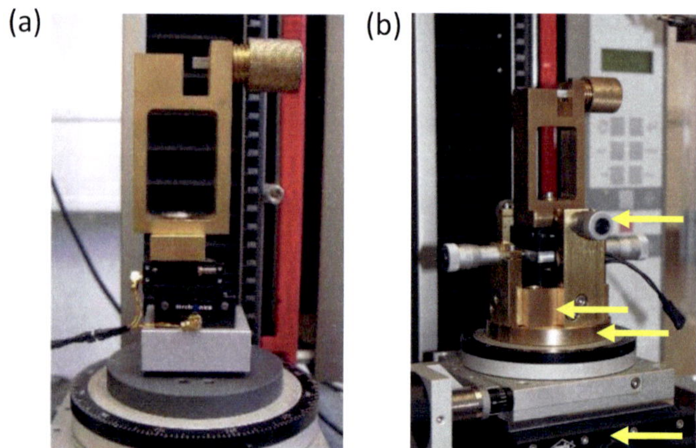

Figure 3.8: Middle part of the test set-up (a) before optimization, (b) after optimization.

Two thick and integrated base plates, as shown in Figure 3.9(a), replaced the old ones which were made from several thin aluminum sheets. The new base plates effectively avoided the vibration of x-z stages during the testing. A large x-y table was added under the rotation table. This made the alignment more precise compared to the situation where only small x-y tables were used.

Figure 3.9(b) displays a conical extrusion which was fabricated with high accuracy. It was fixed exactly in the center of the rotation table. This conical extrusion is used for pre-alignment of the system into the rotation axis.

After pre-alignment, a square hollow block, as shown in Figure 3.9(c), can be directly placed onto the conical extrusion. This process avoided any mismatch during the process of moving away the conical extrusion in the original set-up and made the alignment as precise as possible. Since the small x-y table above the rotation table could be moved mechanically during mounting and dismounting a sample, a misalignment could be introduced. This may significantly influence the noise of the torque curves. Therefore, four helical rods were used to fix the small x-y table to perfect each alignment.

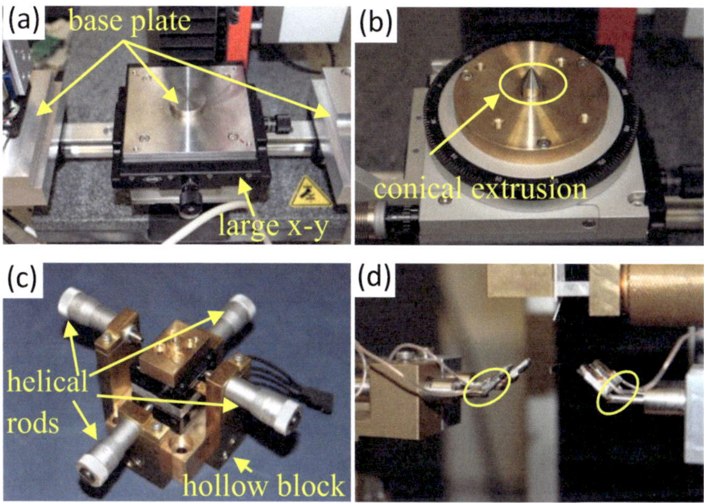

Figure 3.9: Optimization of test set-up: (a) base plate and large x-y table, (b) conical extrusion for pre-alignment, (c) helical rods to fix small x-y tables and (d) the angle of y-positioners.

Moreover, the angle between the load sensors and the cross-beam also has an impact on the conversion factor on the load, which will be described in Section 3.4.2.1. Therefore, the angle of load sensor has to be fixed using super glue (Loctite 401 Ethyl cyanoacrylate), as presented in Figure 3.9(d). This can effectively prevent a user from changing the angle when operating the set-up close to the sensors. Otherwise, a new calibration will be necessary when there is a change in the positions of load sensors.

As mentioned in Section 3.4.1.1, a cross beam is glued onto the torsion specimens. Figure 3.10(a) shows how a cross beam moves around a circular path during a complete rotation in the presence of a misalignment.

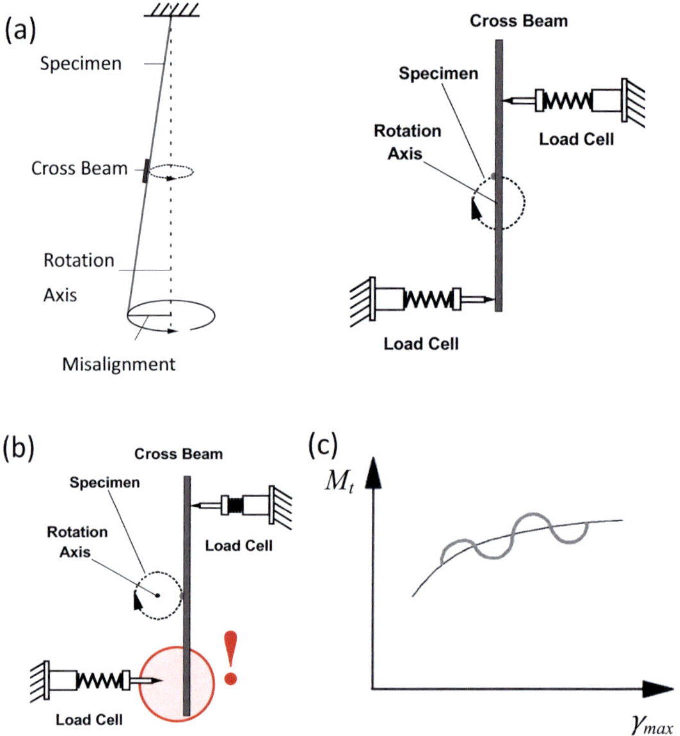

Figure 3.10: Influence of a vertical misalignment of a wire on the measurement: (a) the cross beam moves around a circular path during a complete rotation, leading to an alternating load on the load sensors, (b) one of the load sensors loses contact during the test, (c) oscillating signal caused by misalignment.

The motion of this cross beam leads to the possibility that one of the load sensors may lose contact as demonstrated in Figure 3.10(b). Besides, the circular motion of a cross beam may affect the arm of lever and thus, result in the variation of the force applied onto each load sensor. Hence oscillating signals will be detected in the force measurement, as illustrated in Figure 3.10(c).

As the AFM-tips are highly sensitive, the surface quality of the cross beams has a significant impact on the torque curve, when a sample is not aligned exactly within the rotation axis. A schematic diagram to describe this influence is depicted in Figure 3.11. The roughness of a cross-beam's surface results in some noise and this noise is superimposing the torsion curve from Figure 3.10(c).

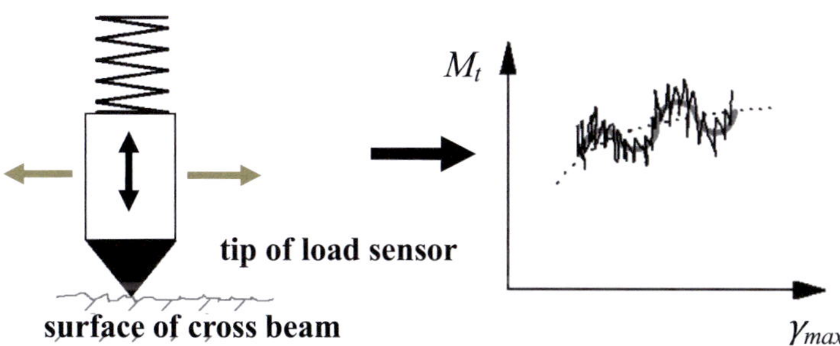

Figure 3.11: Schematic diagram showing the influence of a cross-beam's surface on the torque curve.

In order to minimize the noise, different methods were used to improve the surface of the cross beams, such as mechanical polishing, electro-chemical polishing or coating. The cross beams are made from Ti-foil with a thickness of 125 and 140 µm, respectively. The mechanical polishing was conducted using a SiC paper (grit 4000) and coating of Ti was carried out by evaporation deposition in an ultra-vacuum chamber where vacuum level was set to be 10^{-6} mbar.

Table 3.2 shows the results of light profilometer investigation for four different surface treatment methods: (i) as-received state, (ii) polished with SiC paper, (iii) polished with SiC paper and coated with 100 nm Ti, (iv) polished with SiC paper, coated with 100 nm Ti and annealed at 750°C for 30 min [162-164]. R_a is the average roughness, which reflects the topography of the surface. R_{max} is the measure of maximum roughness. It is the distance, in height, measured from the top of the highest peak to the bottom of the lowest valley of a surface topography. According to Table 3.2, it was found that the surface quality was improved by mechanical polishing and coating. The results reveal that the 125 μm thick cross beam has a better surface quality than the 140 μm thick cross beam, especially after coating. Therefore, further investigations were conducted on the cross beam with a thickness of 125 μm.

Table 3.2: The roughness of different surface treated cross-beams made from Ti

	140 μm thick cross beam		125 μm thick cross beam	
	R_a (μm)	R_{max} (μm)	R_a (μm)	R_{max} (μm)
(i) as-received state	0.52	5.32	0.20	1.69
(ii) polished with SiC paper	0.19	2.92	0.13	1.99
(iii) polished with SiC paper and coated with 100 nm Ti	0.139	2.457	0.098	1.007
(iv) polished with SiC paper, coated with 100 nm Ti and annealed at 750°C for 30min	0.21	3.34	0.13	2.48

These cross beams were coated with 200 nm Ti and annealed at 500°C for 30 min. Table 3.3 lists the surface roughness determined by atomic force microscopy (AFM Multimode Nanoscope III, Digital Instruments-Veeco). The results reveal that the surface quality after coating and annealing was significantly improved compared to the original surface quality of cross beams. Based on these investigations, three types of cross beams were produced from the 125 μm thick Ti foil with the size: 22 mm × 2 mm, 32 mm × 1.5 mm, 45 mm × 1.5 mm. Length of cross beams was varied to adjust the load corresponding to the torsional moment.

Table 3.3: The surface roughness of the 125 μm thick cross beams determined by atomic force microscopy (AFM)

	R_a (μm)	R_{max} (μm)
as-received state	0.21	1.95
coating 200 nm Ti and annealed at 500°C for 30 min	0.007	0.98

3.4.2   Experimental Procedures

3.4.2.1   Alignment and calibration

As described in the previous sub-chapter, the alignment of the wires with the rotation axis of the rotation table is very important. Figure 3.12 displays the key steps for alignment. The alignment in the horizontal direction is done prior to the vertical direction. A high resolution water level device with a resolution of 20 μm on 1 m is used for the horizontal alignment of the test-up, as shown in Figure 3.12(a).

Figures 3.12(b) and (c) illustrate the vertical alignment along the rotation axis. First, a pre-alignment is conducted using a plummet clamped in the upper specimen grip to make the tip of plummet as close as possible to the conical extrusion by adjusting the large x-y table. This step is used to align the sample to the center of the rotation table. After mounting the middle part (x-y table, helical rods and lower specimen grip), the final vertical alignment of the test set-up should be performed. The plummet is again used to align the specimen along the rotation axis by adjusting the small x-y table. This step can eliminate the mismatch among different components in the middle part of the test set-up.

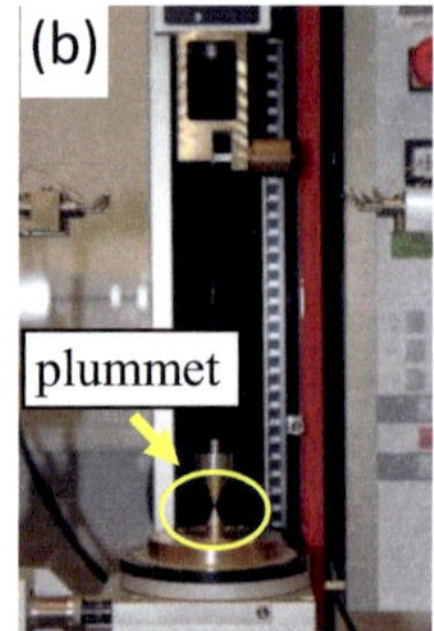

Figure 3.12: Alignment of the test set-up: (a) horizontal alignment using a water level device, (b-c) vertical alignment using a plummet.

As mentioned in Section 3.4.2.3 (refer to Figure 3.9(d)), the measured load is depending on the tilt angle of the load sensors. The conversion factor of the load for each individual load sensor should be determined. A tungsten wire with a diameter of 50 µm (Goodfellow, Germany) is used as a reference sample to calibrate the load sensors, since tungsten is an isotropic material with a well-known shear modulus. With the given geometrical values of a specimen, the theoretical load F [μN] can be calculated as a function of the torque angle:

$$F[\mu N] = \frac{GI_P}{la}\varphi \qquad (3.4)$$

with the shear modulus G, the length of the specimen l, the polar moment of inertia I_p, the arm of lever a, and the torque angle φ, which is given by the rotation table. The polar moment of inertia I_p for wires / cylinders is: $I_P = \frac{\pi}{32}d_{wire}^4$.

The relationship between the load in voltage obtained from the measurement and the load in Newton, is expressed as:

$$F[\mu N] = m \cdot F[V] \qquad (3.5)$$

where the conversion factor m can be calculated from relationship between $F\,[\mu N]$ and $F\,[V]$, as shown in Figure 3.13.

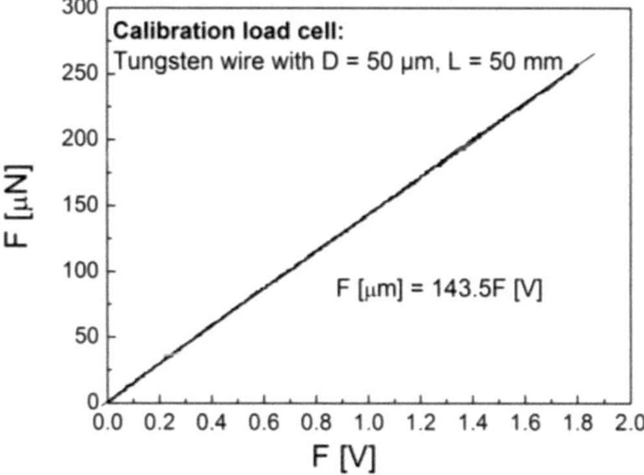

Figure 3.13: Calibration curve of a load sensor based on a torsion test on a tungsten wire with a diameter of 50 µm. $F\,[V]$ is the load in voltage obtained from the measuring system and $F\,[\mu N]$ is the calculated load in Newton.

3.4.2.2 Sample preparation and test conditions

Figure 3.14 shows the preparation of samples used for the mechanical tests. The specimens for both tensile and torsion tests are bonded onto a paperboard frame, as shown in Figure 3.14(a), to avoid any pre-deformation or damage of the wires. The width of the frames is precisely adapted according to the size of the notches in the specimen grips. A metallic positioning device with a sharp V-notch (refer to Figure 3.14(b)), is used to prepare the specimens. During sample preparation, misalignment can be controlled within 10 μm relative to the reference side of the frame with the aid of the positioning device. For torsion tests, a cross beam is additionally bonded to the wire using a stiff super glue (Loctite 401 Ethyl cyanoacrylate). The initial gauge length of the tensile and torsion specimens is set to be $l_0 = 30$ mm and $l_0 = 50$ mm, respectively.

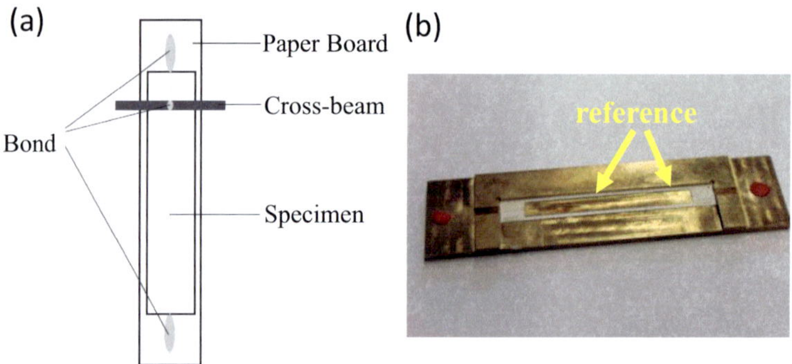

Figure 3.14: (a) Schematic illustrations of a prepared specimen for torsion test, (b) metallic positioning device with a V-notch.

The tensile tests were conducted under displacement control at displacement rates of 0.001, 0.01 and 0.1 mm/s. The unloading segments of the tensile tests were used to determine the Young's

Modulus of the wires, since the modulus is an indication of texture differences within the different heat treated wires. The loading segments were performed under displacement control with a displacement rate of 0.01 mm/s, whereas the unloading segments were performed under load control with an unloading rate of 10 mN/s. Torsion tests were carried out at angular velocities of $d\varphi/dt$ = 0.086, 0.172 and 0.345 rad/s.

3.4.2.3   Data analysis

As one can see in Figure 3.15, the determined torque curves show oscillating signals due to the misalignment of the set-up and the sample preparation, as well as the surface roughness of the cross beams. Consequently, torsion curves with similar trend may partly overlap each other, which make comparison of several results in one graph difficult. To determine and to visualize small differences between different moment - shear strain curves, the Ramberg-Osgood law [165]:

$$\varepsilon = \frac{\sigma}{E} + K \left(\frac{\sigma}{E} \right)^n \tag{3.6}$$

expressed for shear in torsion [160]:

$$\gamma_{max} = \frac{M_t}{G \cdot W_P} + \alpha \cdot \frac{M_{t0}}{G \cdot W_P} \cdot \left(\frac{M_t}{M_{t0}} \right)^n \tag{3.7}$$

was used to fit the curves. Here, $\gamma_{max,}$ is the maximum shear stress at the surface, $W_P = \dfrac{\pi}{32} d_{wire}^3$ is the polar moment of inertia, and α, n are fitted material constants.

Figure 3.15 illustrates the validity of this approach, as shown by the results obtained from a wire with $d = 15$ µm, using $G = 42$ GPa, $M_{t0} = 0.0499$ µNm. These values were determined by the calculated torsion curve, and the fitted constants were $\alpha = 3.24$ and $n = 2$. Besides, the hardening exponent n can be used to quantify differences in the hardening behavior of the wires in dependence of the loading rate or the microstructural features.

In general, the M_t torque curves show a 'natural' size effect that bigger is stronger since M_t is proportional to the wire radius R ($M_t = f(I_p) = f(R^4)$). For comparing the torque behavior of different thick wires, normalized torque moments M_t/R^3, representing average shear stresses across the wire radius, are used.

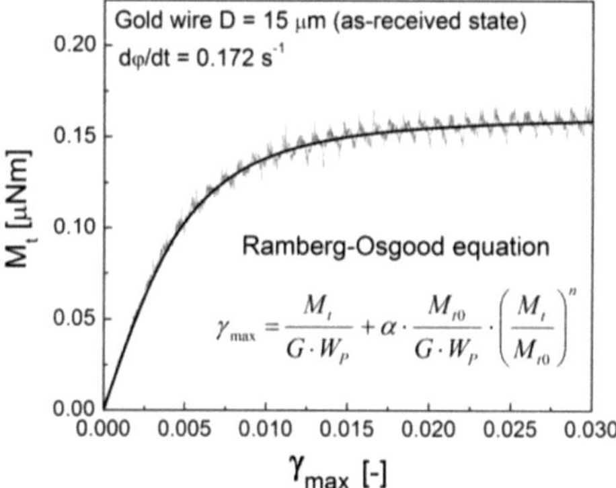

$$\gamma_{max} = \frac{M_t}{G \cdot W_P} + \alpha \cdot \frac{M_{t0}}{G \cdot W_P} \cdot \left(\frac{M_t}{M_{t0}} \right)^{n}$$

Figure 3.15: Fit of the Ramberg–Osgood law to the torque curve of a wire with a diameter of 15 µm.

4. Results

In this chapter, experimental observations and results are presented. Section 4.1 introduces the microstructures and the mechanical properties of different thick gold wires in the as-received state. Results from systematical heat treatment investigations and mechanical tests will be presented in Section 4.2. Investigation on gold wires in two specified annealed states: (1) an intermediate state and (2) a fully recrystallized state, will be outlined in Section 4.3. Section 4.4 presents the distributions of geometrically necessary dislocations in polycrystalline and bamboo-structured twisted gold wires. Finally, the microstructures and the mechanical properties of as-received aluminum wires are presented in Section 4.5.

4.1 Structure and mechanical properties of gold micro-wires in the as-received state

4.1.1 Microstructural characterizations

Figure 4.1 shows cross-sectional micrographs of as-received gold wires with diameters ranging from 15 to 60 µm. The wires reveal fine grained structures with slightly increasing average grain sizes ($0.54 \leq d_g \leq 0.77$ µm) with increasing wire diameter.

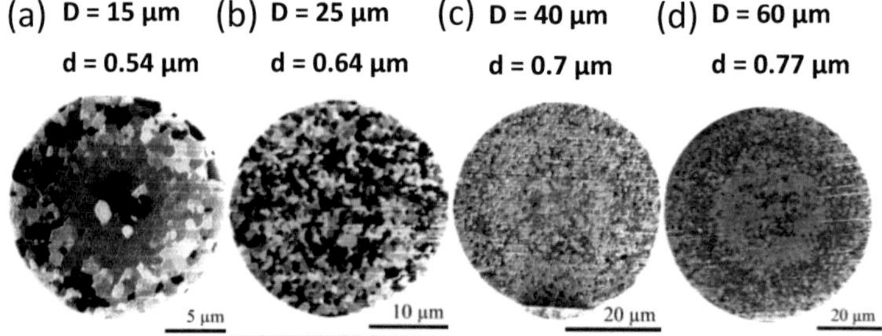

Figure 4.1: FIB micrographs from cross-sections of the as-received gold wires.

Figure 4.2 shows EBSD orientation maps of the gold wires in the as-received state. Textural domains with different colors can be observed in these orientation maps. The center and the surface regions of the wires are predominantly <100> oriented (red color). In contrast, the grains around the center are mainly <111> oriented (blue color). The region between the blue domain and the red layer at the surface exhibits various colors such as violet, light blue, green etc.. The fraction of this isotropic grain section increases with increasing wire diameter. In summary it can be seen that the 25, 40 and 60 µm wires reveal almost isotropic microstructures, whereas the 15 µm wire is more oriented towards the <111> direction.

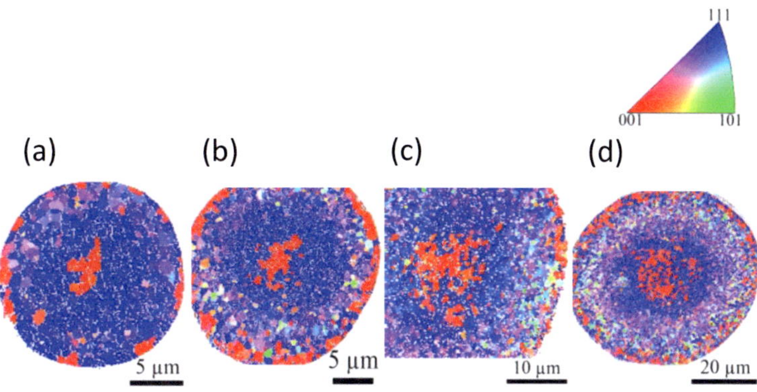

Figure 4.2: Orientation maps determined by electron backscatter diffraction (EBSD) of as-received gold wires with diameters of: (a) 15 µm, (b) 25 µm, (c) 40 µm, and (d) 60 µm.

4.1.2   Deformation behavior in tension and torsion

The effect of strain rate on the tensile behavior of the gold wires in the as-received state is shown in Figure 4.3. The wires show no strain rate sensitivity in the proof stress when displacement rates of 0.001, 0.01 and 0.1 mm/s were applied. However, one can observe that both the uniform elongation ε_u and the elongation at fracture ε_f increase with increasing displacement rate, independent of the wire diameter.

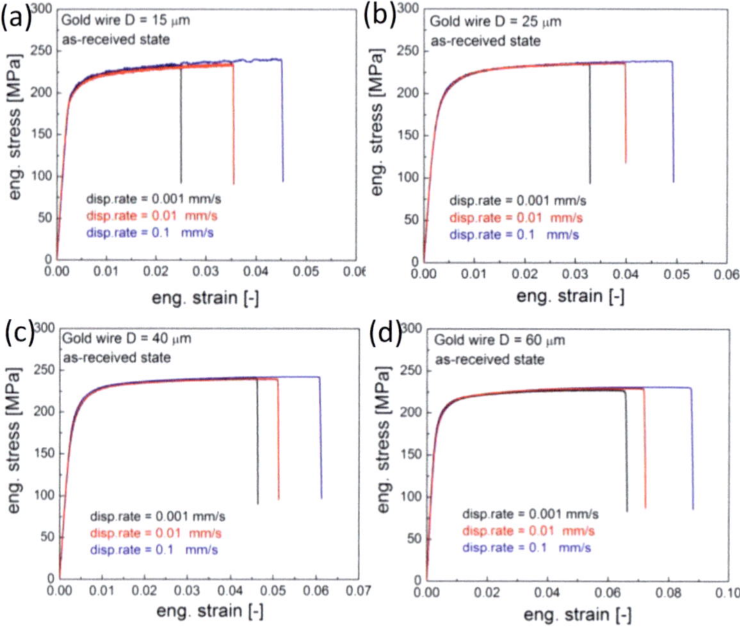

Figure 4.3: Stress-strain behavior of the different thick gold wires in the as-received state under uniaxial loading as a function of the displacement rate.

Figure 4.4 shows the same data, how comparing the influence of the wire diameter at different displacement rates ((a) 0.001, (b) 0.01 mm/s and (c) 0.1 mm/s). It can be seen that the general deformation behavior of the wires is almost size independent. Only the 60 μm wires do show a little lower proof stress and ultimate tensile strength. Besides, it is observed that the uniform elongation ε_u and the elongation at fracture ε_f increase with increasing wire diameter.

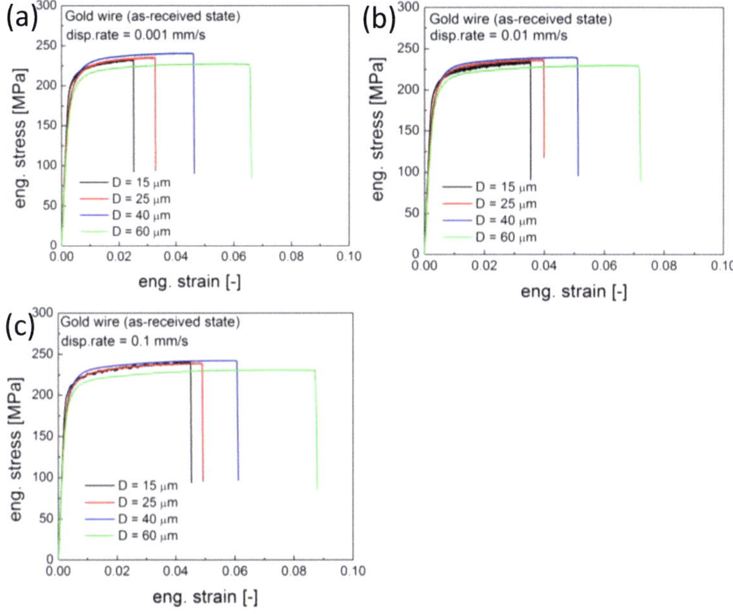

Figure 4.4: Tensile behavior of the as-received gold wires with various diameters loaded at different displacement rates of: (a) 0.001 mm/s, (b) 0.01 mm/s, and (c) 0.1 mm/s.

At the onset of necking, plastic deformation is localized at the 'weakest' section of a specimen. The absolute value of the elongation on necking depends on the ratio between the gauge length (l_0) and the diameter (D) of a sample. For the same wire diameters, an increasing ratio of l_0/D will lead to decreasing values of the calculated elongation ($\Delta l/l_0$). Since the specimens used in this study exhibit very high l_0/D ratios, the uniform elongation ε_u is almost identical with the elongation at fracture ε_f. However, every wire exhibits at least 50 % lower fracture engineering stresses compared to its respective ultimate tensile stress. Accordingly, all tested samples reveal a distinctive necking behavior. For verification, additional SEM investigations on the broken specimens were performed. Figure 4.5 shows the necking zones of tested wires with different diameters in dependence of deformation rate.

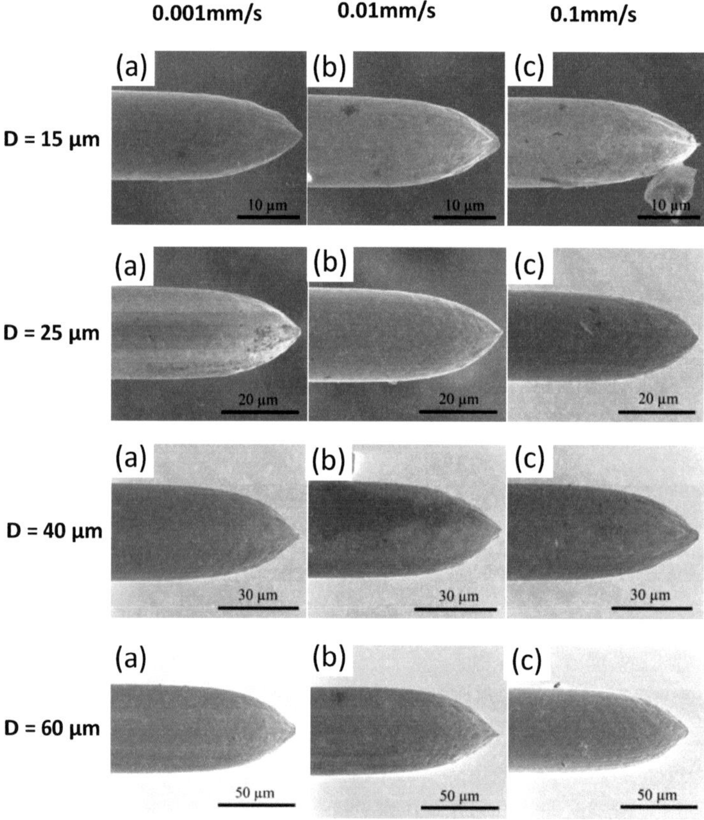

Figure 4.5: Necking zones of tensile tested as-received gold wires in dependence of both wire diameter and applied displacement rate.

One can see that, independent of the wire diameter, the shapes of the zones are almost identical. However, it was found that the length of the necking zones increased slightly with increasing wire diameter.

To quantify the observed textures in the different thick wires (refer to Figure 4.2), additional tensile tests with unloading segments were conducted to determine the Young's modulus of individual samples. Figure 4.6 illustrates the loading profile to determine the Young's modulus, taking a test on the 40 µm wire as an example. *E1* to *E6* correspond to the moduli determined from the 1st to the 6th unloading segment of the true stress-strain curve.

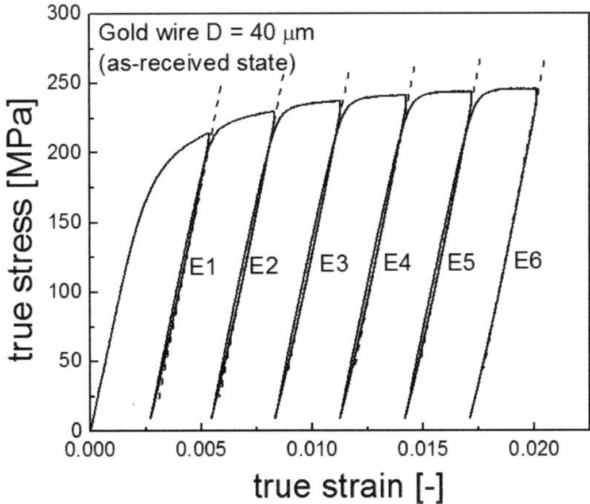

Figure 4.6: Example of a true stress-strain curve with unloading segments in order to determine the Young's modulus of a wire.

The results of all the tests are listed in Table 4.1. The Young's moduli are determined from the average value of the slopes from different unloading segments. When comparing the determined values with the known Young's moduli of bulk gold (E_{pc} = 78 GPa, $E_{<111>}$ = 117 GPa, $E_{<100>}$ = 43 GPa [22]), it was found that the thinner wires show higher Young's moduli. The grains of the 15 µm wires are preferentially oriented towards the <111> direction. In contrast, the 25, 40 and 60 µm wires show almost isotropic behavior.

Table 4.1: The Young's modulus of gold wires in the as-received state determined from tensile tests with unloading segments

Diameter	$E1$ [GPa]	$E2$ [GPa]	$E3$ [GPa]	$E4$ [GPa]	$E5$ [GPa]	$E6$ [GPa]	Average value [GPa]
15 µm	90	90	90	90	90	91	91
	91	92	91	93	92	91	
	91	90	90	90	90	91	
25 µm	82	82	81	82	82	82	82
	82	84	83	82	83	82	
	83	82	82	82	82	84	
40 µm	81	80	80	79.	80	81	81
	82	81	80	81	81	82	
	81	80	81	81	81	81	
60 µm	79	78	80	80	79	79	79
	79	79	79	79	79	79	
	77	77	78	78	77	78	

Figure 4.7 shows the relationship between torque M_t and surface shear strain γ_{max} for wires with different thickness and subjected to different twist rate. It was found that the torque behavior was quite similar when the wires were loaded at angular velocities ($d\varphi/dt$) of 0.017, 0.086, 0.172, 0.345 and 0.862 rad/s. The differences in the torque do not show any systematic trend. Accordingly, one may assume that the torsional responses of these wires are insensitive to the twist rate.

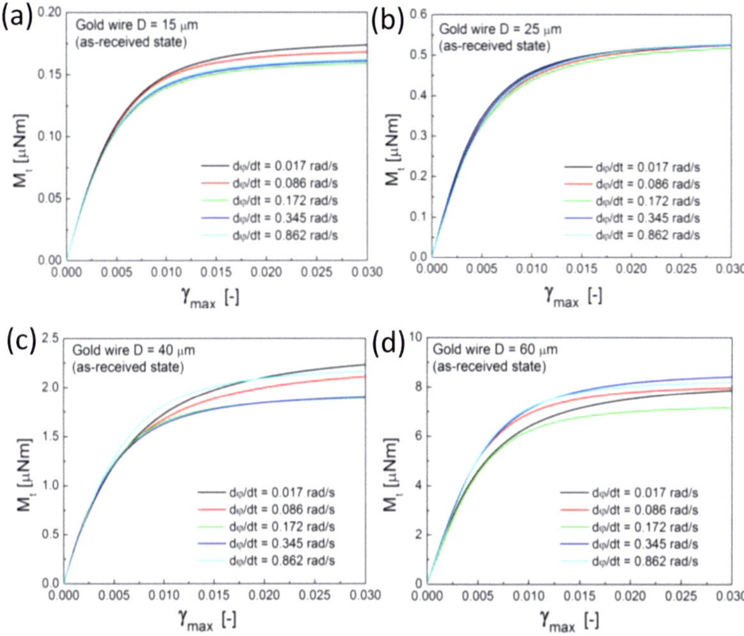

Figure 4.7: Torque behavior of the gold wires in the as-received state with diameters of: (a) 15 μm, (b) 25 μm, (c) 40 μm, and (d) 60 μm as a function of the angular velocity.

Figure 4.8 shows the relationship between normalized torque M_t/R^3 (with $R = D/2$) and surface shear strain γ_{max} (see Section 3.4.2.3). The torque behavior of the gold wires in dependence of the diameter is illustrated at different angular velocities ($d\varphi/dt$) of 0.017, 0.086, 0.172, 0.345 and 0.862 rad/s. In contrast to tension, a significant size effect was observed for the 15 μm wire compared to the 25, 40 and 60 μm wires, which are independent of the deformation rates. The thicker wires show comparable strength similar to the tensile behavior.

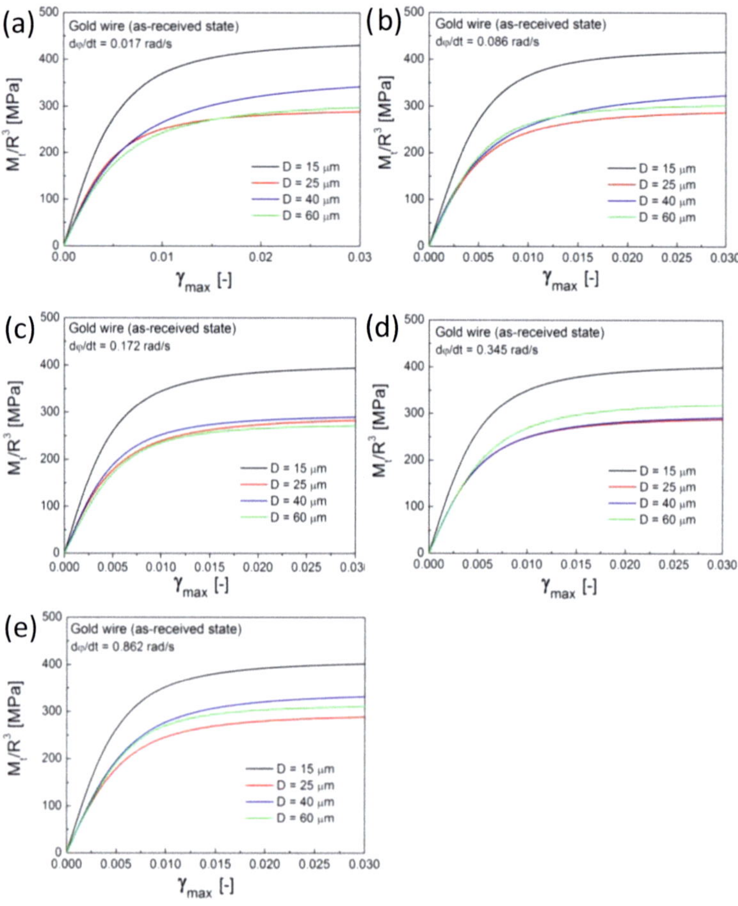

Figure 4.8: Normalized torsional response of the as-received gold wires with various diameters loaded at different angular velocities dφ/dt of: (a) 0.017 rad/s, (b) 0.086 rad/s, (c) 0.172 rad/s, (d) 0.345 rad/s, and (e) 0.862 rad/s.

4.2 Heat treatment investigations

The gold wires with different diameters (15, 25, 40 and 60 µm) were annealed to obtain different grain sizes. As shown in Figure 4.9, the resulting microstructures strongly depend on the annealing temperature. A transition region is observed for each wire diameter. Below this transition region, grain growth is negligible and mainly the recovery processes are active. Above this transition region, the grain size increases rapidly with increasing temperature. The corresponding samples reveal completely recrystallized coarse-grained microstructures. Interestingly, the transition temperature from recovery to recrystallization increases with increasing wire diameter.

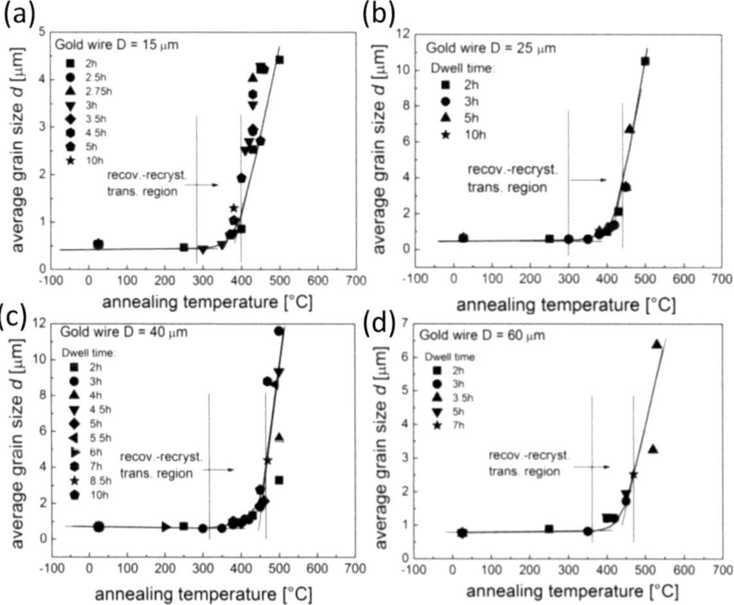

Figure 4.9: Influence of the annealing temperature and dwell time on the average grain size of gold wires with diameters of: (a) 15 µm, (b) 25 µm, (d) 40 µm, and (d) 60 µm.

Besides, it was found that the variation in dwell times at same annealing temperature insignificantly affected the changes in the average grain size. In other words, the annealing temperature has a significantly stronger influence on the microstructural evolution than the dwell time.

The influence of the annealing temperature on the mechanical behavior of the wires was investigated in both tension and torsion. Figure 4.10 takes 15 and 60 µm wires as examples to show the tensile and torsional behavior of gold wires in the as-received state, an intermediate state (annealed at 400°C for 2 hours) and an fully recrystallized state (430°C for 3.5 hours for 15 µm wires and 530°C for 3.5 hours for 60 µm wires).

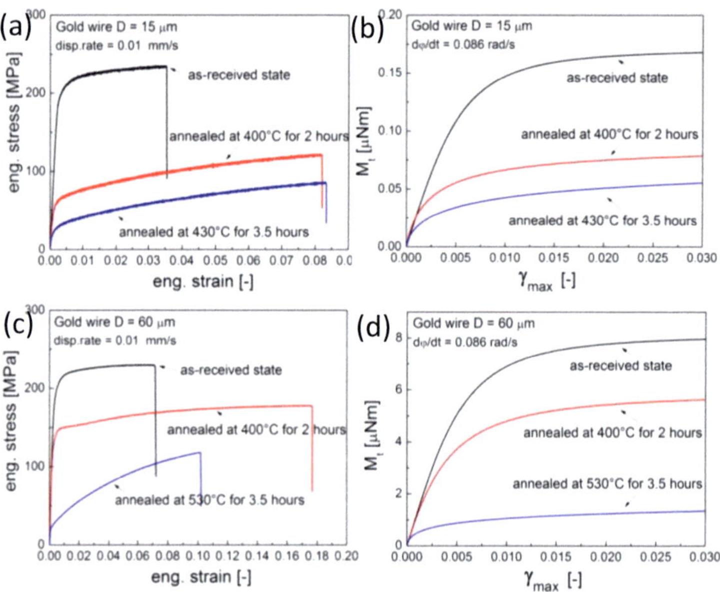

Figure 4.10: Influence of annealing temperature on the tensile and torsion behavior, as exemplified by the 15 and 60 µm wires.

Both tensile and torsional behavior show the trend that the flow stress decreases significantly below 400°C for the 15 μm wire and above 400°C for the 60 μm wire. It was found that the hardening increased with increasing temperature in tension and decreased in torsion. A systematic investigation on the tensile behavior with adjusted annealing temperatures was done. A general relationship between the yield stress and the annealing temperature is established in order to obtain required mechanical behavior for further investigations.

Based on the results obtained from the heat treatment investigations and the mechanical tests, two annealed states were chosen for further investigations in order to study the deformation behavior of gold wires in the presence of strain gradients more detailed:

1. 'Intermediate state': the different thick wires were all annealed at 400°C with a dwell time of 2 hours. In this case, the wires reveal similar average grain sizes of about 1 μm but exhibit different degrees of recovery.

2. 'Fully recrystallized state': individual heat treatments were performed to achieve similar deformation behavior in tension. The heat treatment parameters for the corresponding wire diameters are listed in Table 4.2.

Table 4.2: Heat treatment parameters for the gold wires in the fully recrystallized state in order to obtain comparable deformation behavior in tension

Nominal diameter (μm)	Heat treatments for comparable mechanical behavior
15	430°C for 3.5 hours
25	450°C for 5 hours
40	490°C for 5.5 hours
60	530°C for 3.5 hours

4.3 Structure and mechanical properties of gold micro-wires in two annealed states

4.3.1 Structure and mechanical properties in the intermediate state

4.3.1.1 Microstructural characterizations

Figure 4.11 displays micrographs of gold wires with diameters ranging from 15 to 60 μm after heat treatments at 400°C for 2 hours. The micrographs show inhomogeneous microstructures, where grains located at the surface are coarser than those in the center. This is attributed to the decreasing degree of recrystallization from the surface towards the center within single wires. This effect is more pronounced in the thinner wires, indicating that a larger fraction of material has recrystallized. Nevertheless, all the wires have similar average grain size. The average grain size of 15, 25, 40 and 60 μm wires was measured to be 0.86, 0.99, 0.96 and 1.21 μm, respectively.

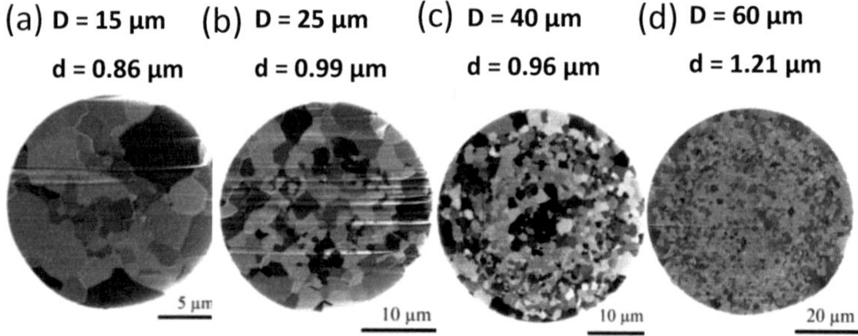

Figure 4.11: FIB micrographs of the cross-sections of gold wires annealed at 400°C for 2 hours.

Figure 4.12 shows orientation maps of these wires in the intermediate state. It was found that the red color indicating <100> direction is mainly distributed near to the surface and the blue color indicating <111> direction is still the most dominant. Compared to the structure in the as-received state, the grains with red color oriented in <100> direction are coarser and occupy more volume fraction. The thinner wire, i.e., 15 µm wire, displays significant grain growth compared to that in the as-received state, which is due to the different degrees of recovery/ recrystallization of the wires. However, the thicker wire, i.e., the 60 µm wire, shows only a small change in the grain size (from 0.77 µm to 1.21 µm).

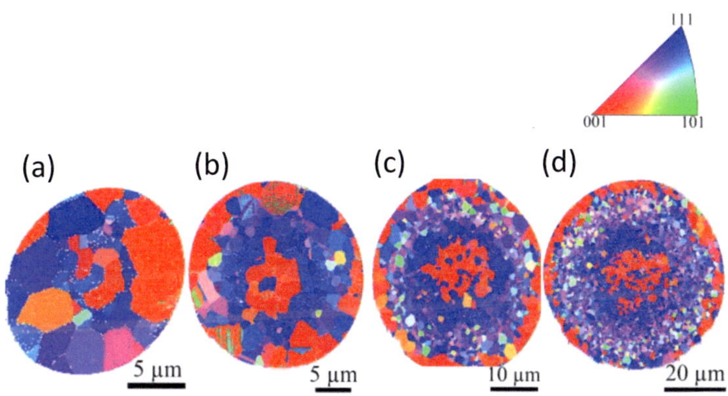

Figure 4.12: Orientation maps determined by electron backscatter diffraction (EBSD) of gold wires in the intermediate state with diameters of: (a) 15 µm, (b) 25 µm, (c) 40 µm, and (d) 60 µm.

4.3.1.2 Deformation behavior in tension and torsion

Figure 4.13 shows the influence of the displacement rate on the tensile behavior of the gold wires in the intermediate state. The hardening of the wires increases slightly with increasing displacement rate. However, the wires exhibit no evident displacement rate sensitivity on the proof stress when the displacement rate increases from 0.001 mm/s to 0.1 mm/s. Moreover, the uniform elongation ε_u and the elongation at fracture ε_f were found to increase with increasing displacement rate. There is only one exceptional case, which was observed in the 60 µm wires.

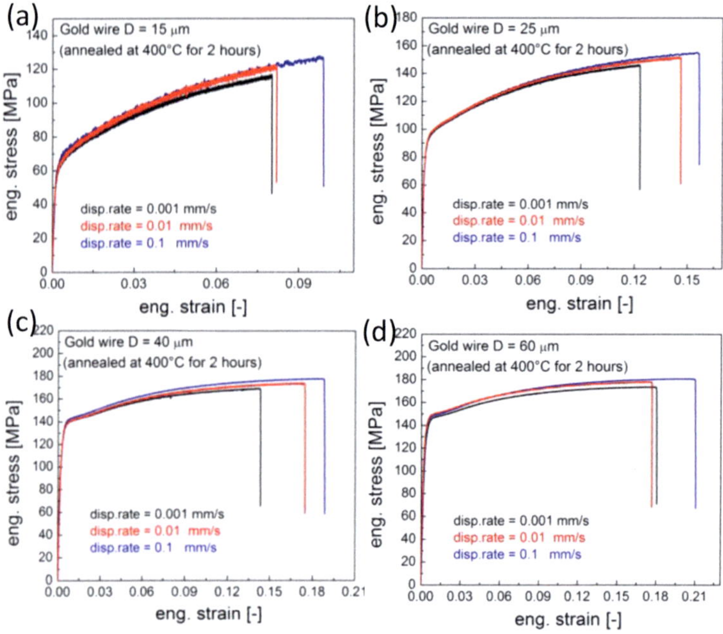

Figure 4.13: The tensile stress-strain behavior as a function of displacement rate of the gold wires in the intermediate state with diameters of: (a) 15 µm, (b) 25 µm, (c) 40 µm, and (d) 60 µm.

Figure 4.14 shows the tensile behavior of different thick gold wires at displacement rates of 0.001, 0.01 and 0.1 mm/s. Based on the degrees of recrystallization, one can see that the strength decreases with decreasing wire diameter. The stress-strain curves show different mechanical behavior of the wires in the region from pure elastic to plastic deformation. The 15 µm wire shows smooth transition in the elastic-plastic region, which indicates continuous hardening after yielding. The 25 µm wire presents the similar behavior as the 15 µm wire. However, the 40 and 60 µm wires show strong hardening at the beginning followed by an evident plateau after yielding, and then with low hardening up to almost ideal plastic behavior. This general trend is thought to be associated with the different degrees of recovery/recrystallization. Furthermore, the uniform elongation ε_u and the elongation at fracture ε_f were found to increase with increasing wire diameter.

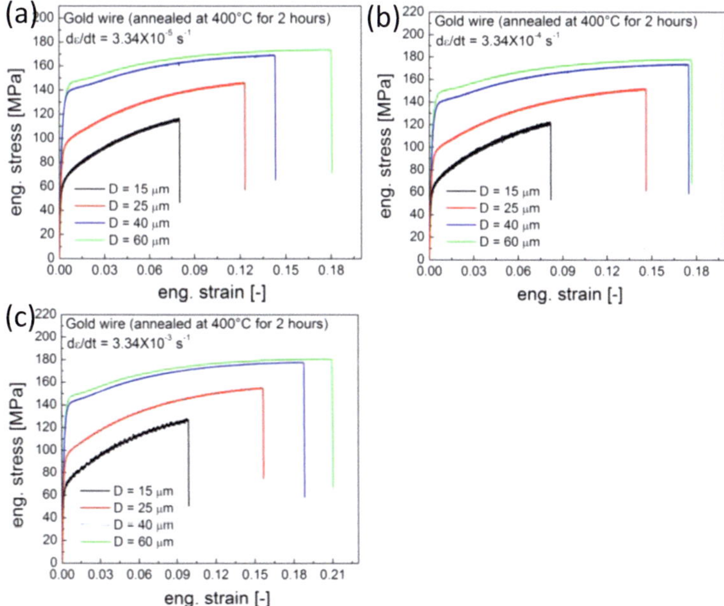

Figure 4.14: Tensile stress-strain behavior of gold wires with various diameters in the intermediate state loaded at displacement rates of (a) 0.001 mm/s, (b) 0.01 mm/s, and (c) 0.1 mm/s.

Table 4.3 compares the Young's moduli of the gold wires in the intermediate state and in the as-received state. The moduli determined from the tensile unloading segments for the wires in the intermediate state are listed in Appendix 1. The Young's modulus of the 15 μm wire is 66 GPa, indicating that the grains are more oriented in the <100> direction. The 25, 40 and 60 μm wires have an approximate modulus of 72 GPa, which demonstrates more isotropic behavior. Compared to the moduli of the wires in the as-received state, the orientation of the 15 μm wire changes significant from <111> towards <100> direction, whereas the orientation of other wires changes slightly. This trend is consistent with the EBSD observation that the 15 μm wire presents evident grain growth. Besides, more red color regions indicating <100> direction appear in the annealed wires compared to wires in the as-received state.

Table 4.3: The Young's moduli of gold wires in the intermediate state compared to the moduli in the as-received state

	$E_{D\,=\,15\,\mu m}$	$E_{D\,=\,25\,\mu m}$	$E_{D\,=\,40\,\mu m}$	$E_{D\,=\,60\,\mu m}$
intermediate state	**66 GPa**	**72 GPa**	**72 GPa**	**71 GPa**
as-received state	92 GPa	82 GPa	81 GPa	79 GPa

Figure 4.15 depicts the relationship of the normalized torque behavior and the wire diameter loaded at an angular velocity $d\varphi/dt$ of 0.086 rad/s. As observed in tension, the 40 and 60 μm wires also show the highest strength in torsion. However, in contrast to the tensile behavior, the 15 μm wire exhibits similar strength to the 40 μm and 60 μm wires and is significantly stronger than the 25 μm wire.

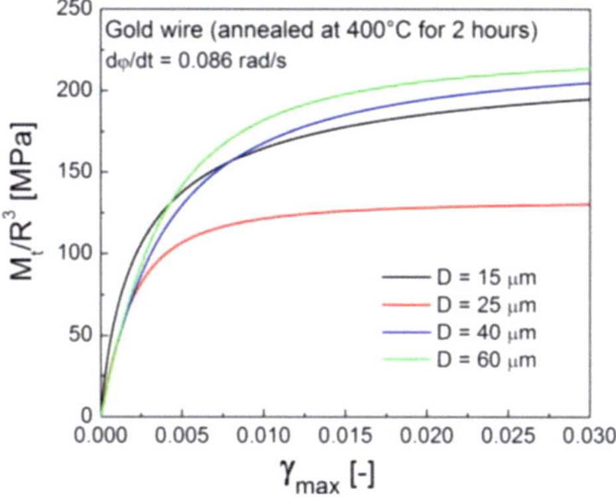

Figure 4.15: Normalized torsional response of different thick gold wires in the intermediate state loaded at an angular velocity $d\varphi/dt = 0.086$ rad/s.

4.3.2 Structure and mechanical properties in the fully recrystallized state

4.3.2.1 Microstructural characterizations

The wires in the fully recrystallized state with comparable tensile behavior at low strains ($\varepsilon < 0.015$, i.e., where both the yielding and the hardening can be detected) were further investigated. Resulting from the individual heat treatments, all wires reveal a coarse-grained structure, as illustrated in Figure 4.16. Comparable microstructures were observed in the 15, 25 and 40 μm wires. These wires have different grain sizes but similar number of grains within a cross-section. In contrast, the 60 μm wire exhibits an inhomogeneous microstructure with coarse grains at the surface and significantly finer grains in the center of the wire.

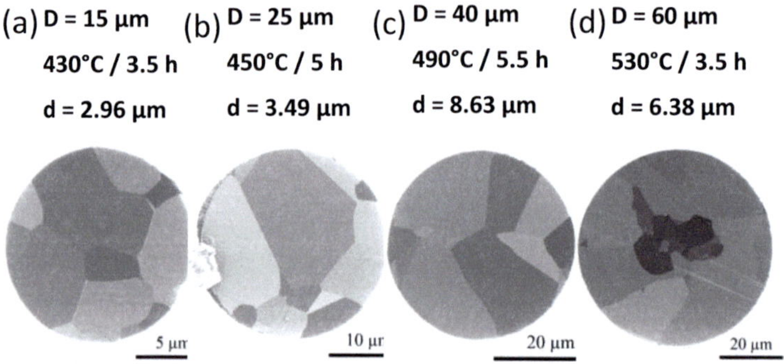

Figure 4.16: FIB micrographs of the cross-sections of gold wires in the fully recrystallized state.

Figure 4.17 shows the orientation maps of these gold wires. The red and blue colors indicating the <100> and <111> directions, respectively, are homogeneously distributed over the cross-sections. A distribution of different grain orientations from the surface to the center of wires is not observed.

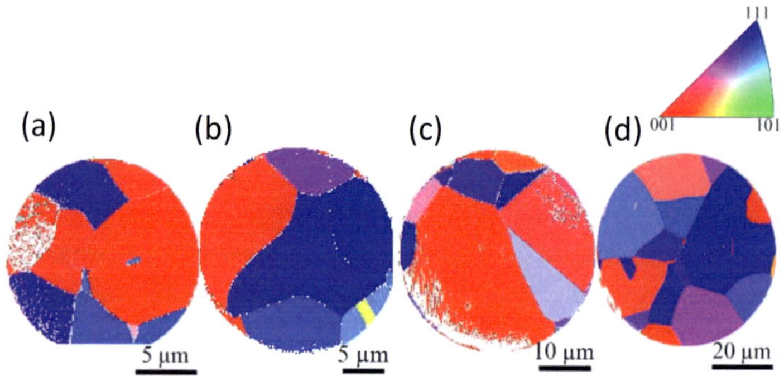

Figure 4.17: Orientation maps determined by electron backscatter diffraction (EBSD) of the gold wires in the fully recrystallized state with diameters of: (a) 15 µm, (b) 25 µm, (c) 40 µm, and (d) 60 µm.

The fraction of the <100> fraction of these wires increases compared to those in the as-received and in the intermediate states, where the red domain is distributed near the surface and in the center of the wires.

4.3.2.2 Deformation behavior in tension and torsion

Figure 4.18 shows the stress-strain behavior of gold wires in the fully recrystallized state in tension as a function of displacement rate. The stress-strain curves reveal smooth transitions from pure elastic behavior towards elastic-plastic behavior. A significant displacement rate dependency was not found.

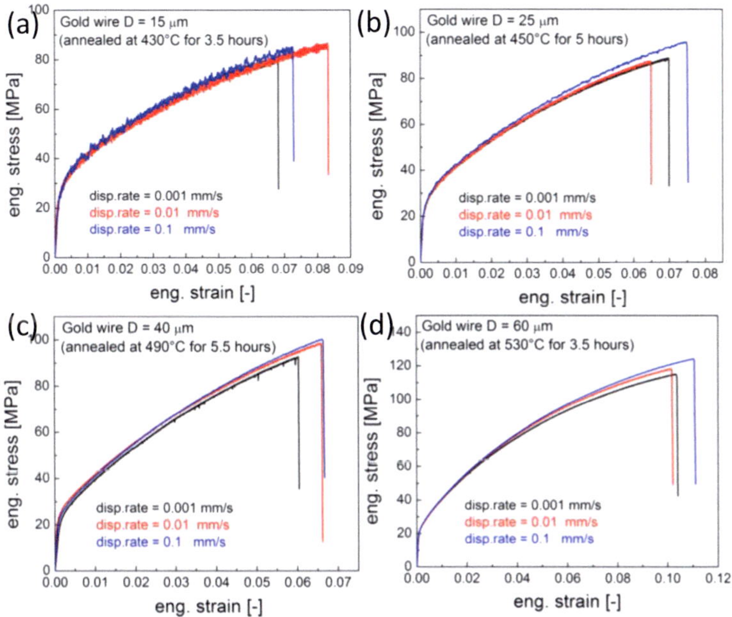

Figure 4.18: The tensile stress-strain behavior of the gold wires in the fully recrystallized state with diameters of: (a) 15 μm, (b) 25 μm, (c) 40 μm, and (d) 60 μm as a function of displacement rate.

However, a slight trend towards increasing hardening with increasing displacement rate was observed. Moreover, the uniform elongation ε_u and the elongation at fracture ε_f for these wires do not show any systematic trend towards increasing values with increasing displacement rate.

Figures 4.19(a)-(c) show the tensile stress-strain behavior of the fully recrystallized gold wires with various diameters at displacement rates of 0.001, 0.01 and 0.1 mm/s. The stress-strain curves reveal comparable deformation behaviors in all wires up to an approximate strain of 0.015. For larger strains, it was generally observed that hardening increases with increasing wire diameter, although the deformation behavior of the 40 and 60 μm wires is quite similar.

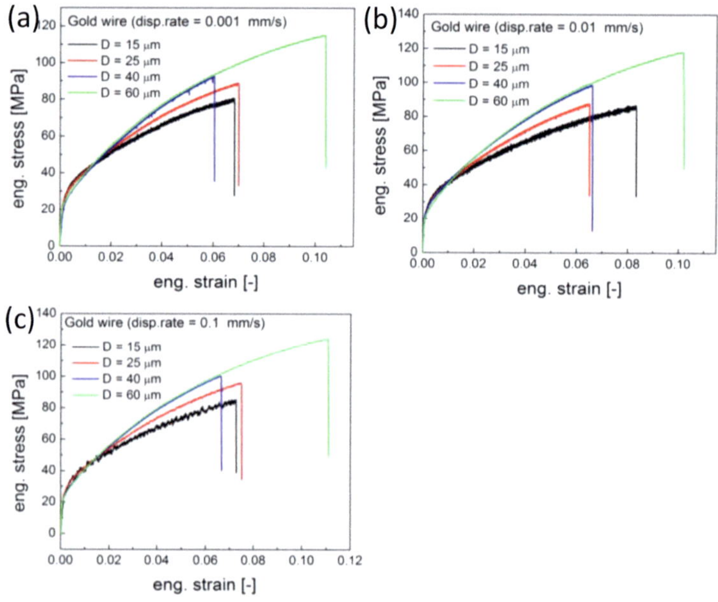

Figure 4.19: The tensile stress-strain behavior of the different thick gold wires in the fully recrystallized state loaded at displacement rates of: (a) 0.001 mm/s, (b) 0.01 mm/s, and (c) 0.1 mm/s.

Table 4.4 compares the Young's moduli of the gold wires in the fully recrystallized state with the moduli in the as-received state and in the intermediate state. The moduli determined from the tensile unloading segments for the wires in the fully recrystallized state are listed in Appendix 2-a. The Young's modulus increases from 51 GPa to 75 GPa as the wire diameter increases from 15 to 60 µm. The 15 µm wire has a modulus of 51 GPa, which is preferentially oriented in the <100> direction, while the 40 and 60 µm wires tend to have more isotropic property. The moduli indicate that the grains are increasingly oriented towards <100> direction in the fully recrystallize state with decreasing wire diameter. Compared to that in the as-received and in the intermediate state, the thinner wire shows larger change in the moduli. This was observed clearly from the evolution of the orientation maps.

Table 4.4: The Young's moduli of gold wires in the fully recrystallized state compared to the moduli in the as-received and in the intermediate state

	$E_{D\,=\,15\,\mu m}$	$E_{D\,=\,25\,\mu m}$	$E_{D\,=\,40\,\mu m}$	$E_{D\,=\,60\,\mu m}$
Fully recrystallized state	**51 GPa**	**64 GPa**	**75 GPa**	**73 GPa**
intermediate state	66 GPa	72 GPa	72 GPa	71 GPa
as-received state	92 GPa	82 GPa	81 GPa	79 GPa

Figure 4.20 shows the influence of the twist rate on the torque behavior of gold wires with different diameters and angular rates. It was found that the torque behavior was similar when wires were subjected to angular velocities ($d\varphi/dt$) of 0.086, 0.345 and 0.862 rad/s. Only the 60 µm wires exhibit some variation in the torque behavior. It seems that the wires tested at low angular velocity ($d\varphi/dt = 0.086$ rad/s) show higher strength. Nevertheless, a systematic trend between the torque and the angular velocity was not observed. This implies that the torque behavior is insensitive to the twist rate in these gold wires.

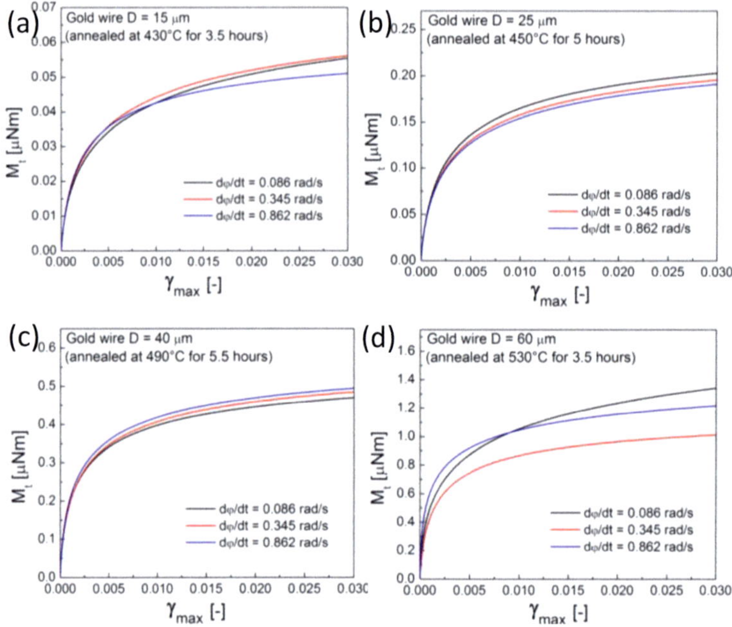

Figure 4.20: Torque behavior as a function of angular velocity for the gold wires in the fully recrystallized state with diameters of: (a) 15 μm, (b) 25 μm, (c) 40 μm and (d) 60 μm.

The normalized torsional response of the gold wires with various diameters loaded at different angular velocities ($d\varphi/dt$ = 0.086, 0.345 and 0.862 rad/s) is illustrated in Figure 4.21. Independent of the angular rate, one can see a distinctive systematic size effect in the sense that 'smaller is stronger'. For larger strains, it is more apparent that hardening increases with decreasing wire diameter.

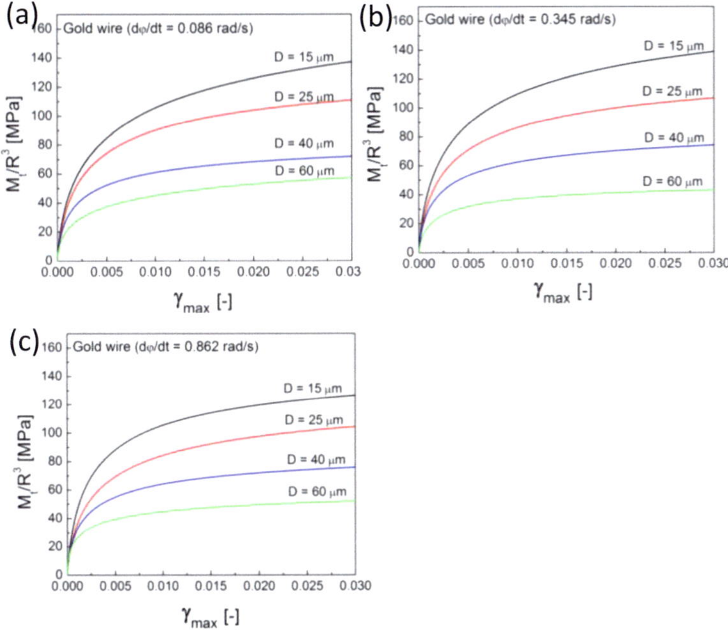

Figure 4.21: Normalized torsional response of different thick gold wires in the fully recrystallized state loaded at different angular velocities $d\varphi/dt$ of: (a) 0.086 rad/s, (b) 0.345 rad/s, and (c) 0.862 rad/s.

4.3.3 Validation of the size effects

4.3.3.1 Microstructural characterizations

In order to validate the observed systematic size effect of the fully recrystallized wires, additional wires with two diameters of 12.5 and 17.5 μm were investigated. Figure 4.22 shows the systematically performed heat treatments on these wires. It was found that the recovery / recrystallization transition temperature for the 12.5 μm wire is lower than that of 17.5 μm wire. There is no obvious change in the grain size for both wires below the transition region, while the grain size increases rapidly with increasing annealing temperature above the transition region.

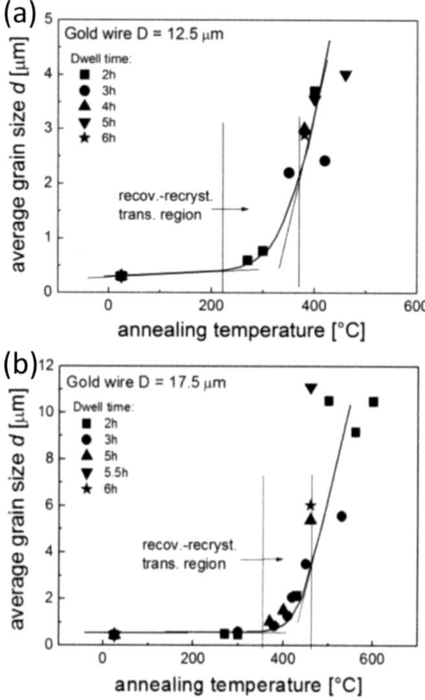

Figure 4.22: Influence of annealing temperature on the average grain size of gold wires with different dwell time with diameters of: (a) 12.5 µm, (b) 17.5 µm.

The annealing parameters were determined to achieve tensile behavior at lower strains (< 0.015) that is similar to the other recrystallized wires. The 12.5 µm wire was annealed at 380°C for 2.5 hours and the 17.5 µm wire was annealed at 460°C for 6 hours. Figure 4.23 shows micrographs of the targeted heat treated 12.5 and 17.5 µm wires. Both wires reveal a relatively coarse-grained structure.

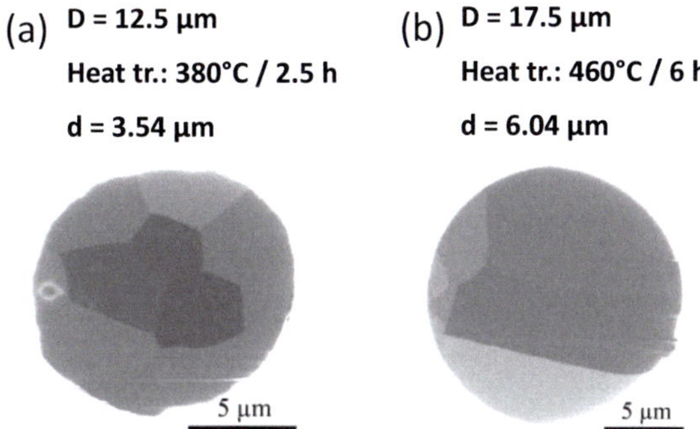

Figure 4.23: FIB micrographs of the cross-sections of gold wires with diameters of 12.5 and 17.5 µm in the fully recrystallized state.

Figure 4.24 shows the orientation maps of these wires in the fully recrystallized state. The single grains are mainly oriented towards <100> and <111> directions, which correspond to red and blue colors in the EBSD images, respectively.

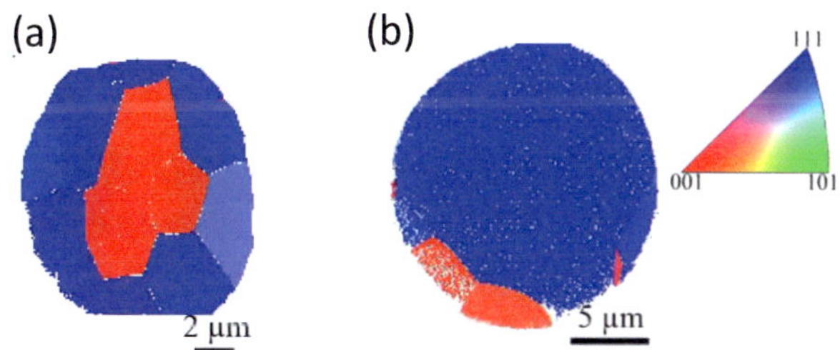

Figure 4.24: Orientation maps determined by electron backscatter diffraction (EBSD) of the gold wires in the fully recrystallized state with diameters of: (a) 12.5 µm and (b) 17.5 µm.

The cross-section of the 17.5 µm wire in Figure 4.24(b) shows dominance in the <111> direction. Depending on the extremely coarse grains of this wire, the dominant orientation may vary along the entire length. This was found in the microstructure of different cross-sections along the longitudinal axis of the wire. Compared to the 15, 25, 40 and 60 µm fully recrystallized wires, the 12.5 and 17.5 µm wires show coarser grains.

4.3.3.2 Deformation behavior in tension and torsion

Figure 4.25 shows the stress-strain behavior of these wires in tension. The stress-strain curves present smooth transitions from pure elastic behavior towards elastic-plastic behavior. There is no evident observation of displacement rate dependency on the flow stress and hardening. However, in contrast to the other fully recrystallized wires, the uniform elongation ε_u and the elongation at fracture ε_f were found to increase with increasing displacement rate.

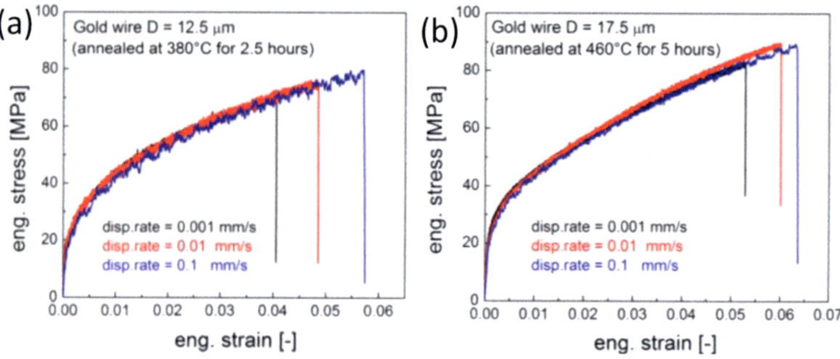

Figure 4.25: The tensile stress-strain behavior of the gold wires in the fully recrystallized state as a function of displacement rate with diameters of: (a) 12.5 µm and (b) 17.5 µm.

Figures 4.26(a)-(c) show the tensile stress-strain behavior of different thick wires at displacement rates of 0.001, 0.01 and 0.1 mm/s. The tensile behavior of the 15 µm fully recrystallized wires is additional depicted for comparison. It can be seen that all wires show similar tensile behavior at lower strains (< 0.015).

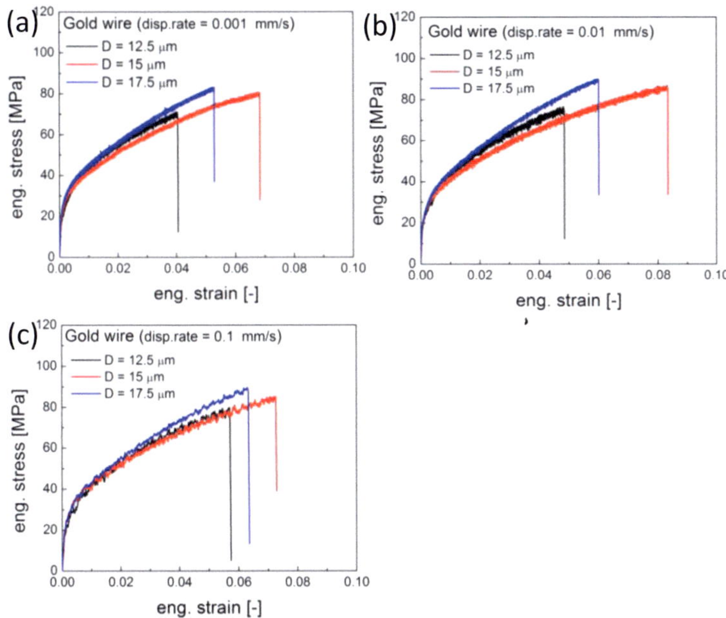

Figure 4.26: The tensile stress-strain behavior of the three thinnest gold wires in the fully recrystallized state loaded at displacement rates of (a) 0.001 mm/s, (b) 0.01 mm/s, and (c) 0.1 mm/s.

Table 4.5 compares the Young's moduli of the 12.5 and 17.5 µm gold wires in the fully recrystallized state compared with the moduli of other different thick wires in this state. The moduli of 12.5 and 17.5 µm wires determined from the tensile unloading segments for the wires in the fully recrystallized state are listed in Appendix 2-b. The Young's moduli for 12.5 and 17.5 µm wires were found to be 65 and 64 GPa, respectively. However, one has to consider that the applied load for 12.5 µm wire was quite low. Thus, the Young's modulus of the single unloading segments could not be determined as accurately. Nevertheless, the moduli of the 12.5 and 17.5 µm wires indicate that the grains of these wires are preferentially oriented towards the <100> direction, which is similar to the preferential orientation of the 15 and 25 µm wires.

Table 4.5: The Young's moduli of gold wires in the fully recrystallized state

$E_{D\,=\,12.5\,\mu m}$	$E_{D\,=\,17.5\,\mu m}$	$E_{D\,=\,15\,\mu m}$	$E_{D\,=\,25\,\mu m}$	$E_{D\,=\,40\,\mu m}$	$E_{D\,=\,60\,\mu m}$
65 GPa	**64 GPa**	51 GPa	64 GPa	75 GPa	73 GPa

Figure 4.27 shows the influence of the twist rate on the torque behavior of these wires. It was found that the torque behavior of the wires is similar when loaded at angular velocities ($d\varphi/dt$) of 0.086, 0.345 and 0.862 rad/s. A systematic trend between the torque behavior and angular velocity is not observed.

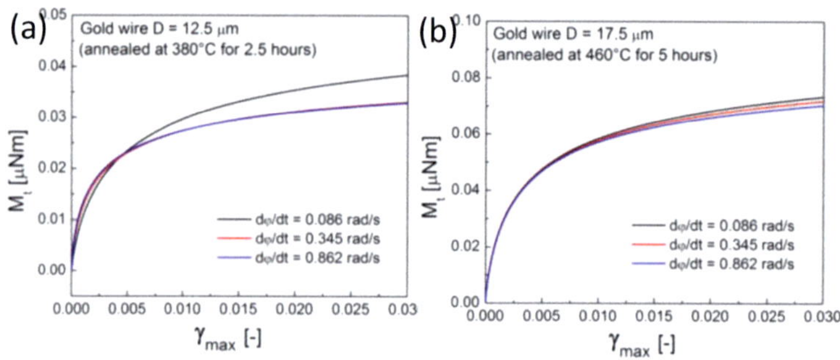

Figure 4.27: Torque behavior of the gold wires in the fully recrystallized state with diameters of: (a) 12.5 µm and (b) 17.5 µm as a function of angular velocity.

The relationship between wire diameter and the normalized torsional response of all the fully recrystallized gold wires loaded at different angular velocities ($d\varphi/dt$ = 0.086, 0.345 and 0.862 rad/s) is illustrated in Figure 4.28. The results of the 12.5 and 17.5 µm wires confirm both the systematical size effect and the increasing hardening behavior with decreasing wire diameter at larger strains.

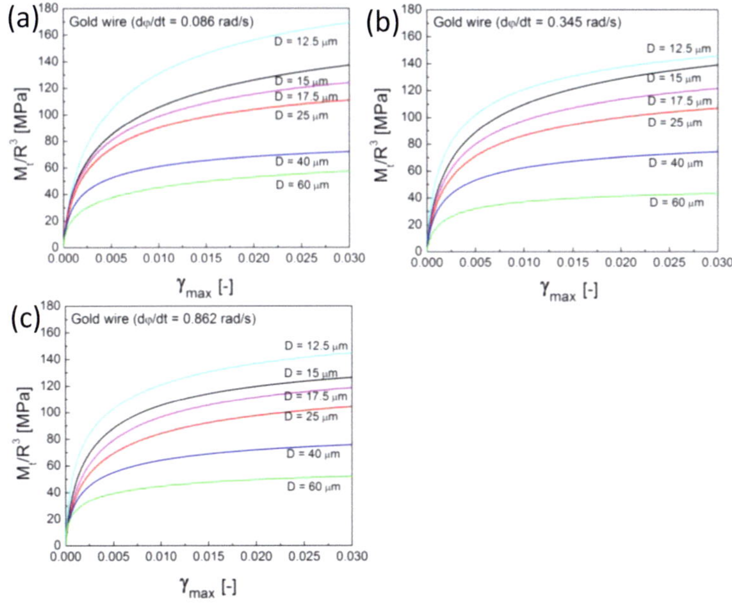

Figure 4.28: Normalized torsional response of all gold wires in the fully recrystallized state loaded at different angular velocities $d\varphi/dt$ of: (a) 0.086 rad/s, (b) 0.345 rad/s, and (c) 0.862 rad/s.

4.4 Ex-situ dislocation structure investigations

In strain gradient theories, geometrically necessary dislocations (GNDs) are assumed to be responsible for the size effect. Under torsional loading, plastic deformation starts at the surface of a sample. It is assumed that the GNDs, resulting from the strain gradient, annularly expand from the surface towards the center of the wires with increasing load in single crystals. However, grain boundaries can play a significant role for the movement of dislocations. Hence, both polycrystalline and bamboo-structured twisted gold wires were investigated in order to determine the GND structure (see Section 3.3.2) at different shear strains.

4.4.1 Dislocation structures in twisted polycrystalline gold wires

Figure 4.29 shows the orientation and misorientation maps and the distribution of GNDs in an annealed and twisted 40 μm wire which was loaded up to a shear strain of $\gamma_{max} = 0.27$. The wire was strongly plastically deformed, which could be confirmed from the inhomogenous distribution of the color within the single grain in the orientation map (Figure 4.29(a)) as well as the misorientation map (Figure 4.29(b)). GND density was calculated based on the misorientation using the (EBSD) method introduced in Section 3.3.2. The distribution of GNDs shows an inverse-star shape pattern, as illustrated in Figure 4.29(c). The highest GND density was found at the surface, and then successively reduced towards the center. The influence of grain boundaries on the GNDs distribution is not evident. This is probably resulting from the low number of grain boundaries in this wire. Figure 4.29(d) shows the average GND density ρ_{gnd} distribution as a function of wire radius r ($0 \leq r \leq R$). Again it is seen that the maximum dislocation density occurs at the surface and decreases towards the center. However, the GND density does not follow a linear behavior. A local peak in dislocation density occurs at a position close to the center of the wire.

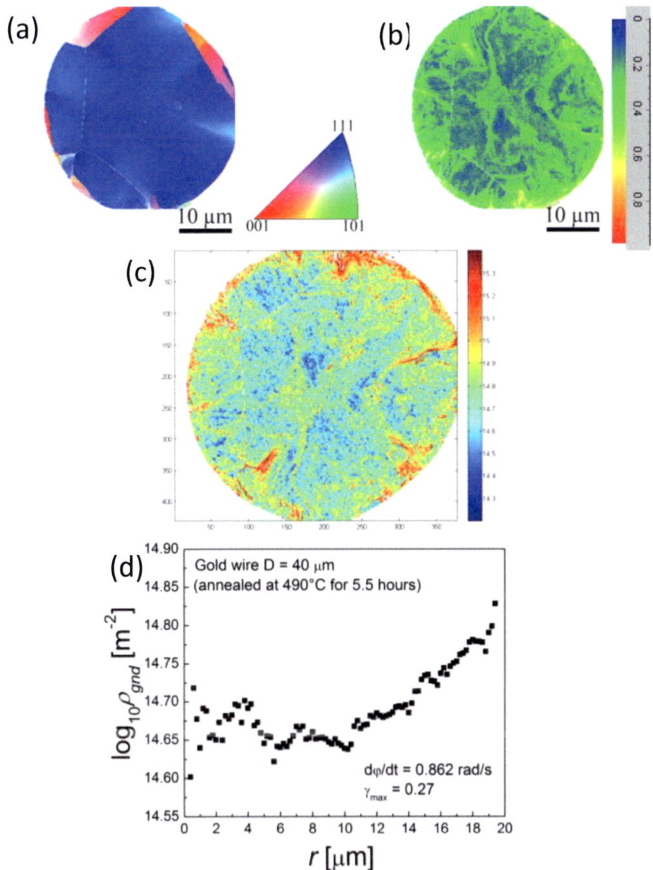

Figure 4.29: Results of EBSD investigations on a twisted 40 µm gold wire (annealed at 490°C for 5.5 hours), loaded up to a shear strain of $\gamma_{max} = 0.27$. (a) Orientation map, (b) misorientation map, (c) distribution of GNDs within the cross-section, (d) distribution of GNDs across the wire radius r. The origin "0" presents the center of the wire.

Figure 4.30 shows the orientation and misorientation maps and the distribution of GNDs in an annealed and twisted 40 µm wire which was loaded up to a smaller shear strain of $\gamma_{max} = 0.08$.

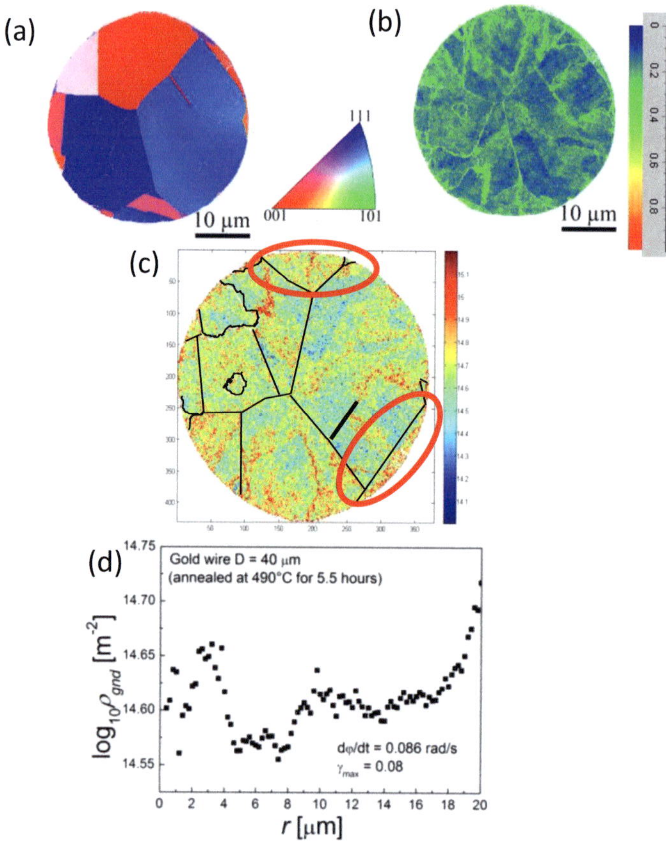

Figure 4.30: Results of EBSD investigations on a twisted 40 μm gold wire (annealed at 490°C for 5.5 hours), loaded up to a shear strain of $\gamma_{max} = 0.08$. (a) Orientation map, (b) misorientation map, (c) distribution of GNDs within the cross-section, where the black lines present the grain boundaries and the red circles show the GNDs restricted by grain boundaries, (d) distribution of GNDs across the wire radius r. The origin "0" presents the center of the wire.

In the misorientation map in Figure 4.30(b), high misorientation regions (bright green) distribute along the grain boundaries or occur in some single grains. Figure 4.30(c) shows the distribution of GNDs within the cross-section. An inverse-star shaped pattern can be observed again. The grain boundaries (black lines) are added to compare the microstructure with the distribution of GNDs. The dislocations were found to be restricted by the grain boundaries, as illustrated in regions marked with red circles. Figure 4.30(d) shows that the GND density ρ_{gnd} decreases from the surface towards the center. However, the decreasing of GND density is not as obvious as the one shown in Figure 4.29(d) for the larger maximum strain. A local peak in dislocation density was also detected at a position close to the center of the wire.

In order to study the influence of grain boundaries on the distribution of the GNDs with increasing shear strain, polycrystalline gold wires with moderate grain size and homogenous structure were investigated. Figure 4.31 shows the distribution of GNDs in annealed and twisted 25 μm wires at different shear strains: γ_{max} = 0.006, 0.039, and 0.16 (corresponding to 1-3 in Figure 4.31). The orientation maps in Figure 4.31(a) show similar grain sizes for all cross-sections. In the misorientation maps in Figure 4.31(b), the structure shows an obvious change in a way that the high misorientation regions (bright green) occur not only in some single grains, as the shear strain increases from 0.039 to 0.16. Figure 4.31(c) displays the distribution of GNDs within the cross-sections. The GNDs are mainly concentrated in grains when the applied strain was small (i.e., γ_{max} = 0.006, 0.039). This implies that grain boundaries have a strong impact on restricting the movement of GNDs. However, there is no systematic relationship between the concentration of GNDs and grain orientations visible when comparing Figure 4.31(a) and (c). The impact of grain boundaries becomes less significant when the shear strain reaches γ_{max} = 0.16. Furthermore, it was found that the surfaces of the wires exhibit higher GND densities. The GND density decreases with increasing distance from the surface to the center, which is consistent with the result shown in Figure 4.31(d-1, shear strain γ_{max} = 0.006). This trend shows good agreement with the strain gradients, with the strain decreasing from the surface to the center of wires. The impact of strain gradients becomes more significant in affecting the distribution of GNDs at large plastic strains. Besides, as seen from the distribution of GNDs across the wire radius r in Figure 4.31(d), the position of the highest GND density was found to move from the surface to the center of the wires as the shear strain increases.

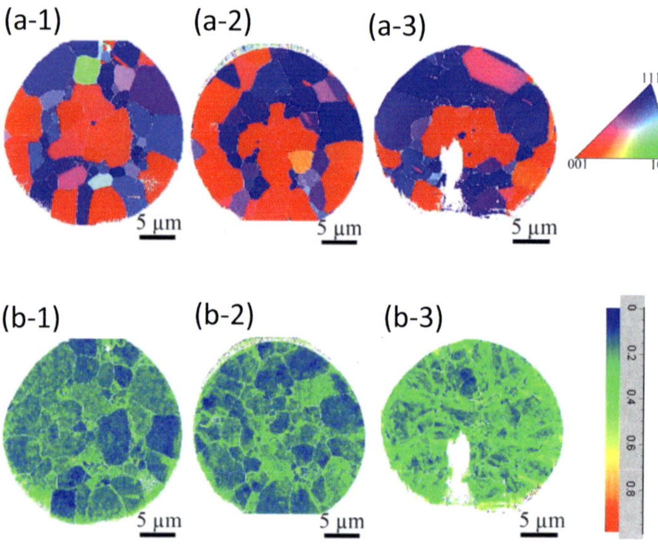

Figure 4.31: Results of EBSD investigations on twisted 25 µm gold wires (annealed at 430°C for 2 hours), loaded at different shear strains of: γ_{max} = 0.006, 0.039, 0.16 (corresponding to 1, 2, 3 in the label, respectively). (a) Orientation maps, (b) misorientation maps

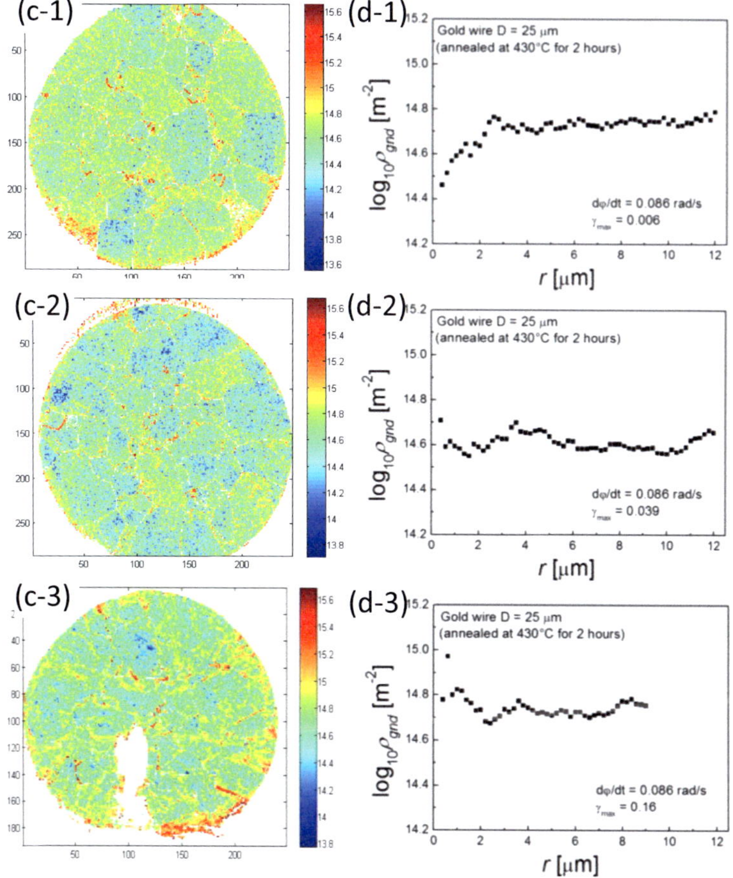

Figure 4.31: Results of EBSD investigations on twisted 25 μm gold wires (annealed at 430°C for 2 hours), loaded at different shear strains of: γ_{max} = 0.006, 0.039, 0.16 (corresponding to 1, 2, 3 in the label, respectively). (c) distribution of GNDs within the cross-sections, (d) distribution of GNDs across the wire radius r. The origin "0" presents the center of the wire.

Coarse-grained fully recrystallized gold wires were also investigated in this study. Figure 4.32 shows the orientation maps and GND distributions of 40 μm wires annealed at 490°C for 5 hours, loaded up to shear strains of: γ_{max} = 0, 0.0014, 0.008, 0.08, 0.16 (corresponding to 1-5 in the figure). The orientation maps of the wires are shown in Figure 4.32(a). Figure 4.32(b) illustrates the evolution of misorientation with increasing shear strain. The misorientation decreases at the beginning and increases for shear strain larger than 0.08. A further increase in shear strain leads to a larger misorientation. The misorientation pattern is significantly distinctive when the shear strain is γ_{max} = 0.16. The high misorientation (bright green) distributes almost over the whole cross-section and the influence of grain boundaries on the misorientation is no longer evident. Figure 4.32(c) shows the distribution of GNDs within the cross-sections. The GNDs are mainly concentrated at the surface and grain boundaries. It was found that the distribution of GNDs was neither solely along the grain boundaries nor annularly from the surface towards the center of the wires at large strains. This is hypothesized to be a coupled effect from both grain boundaries and strain gradients. Figure 4.32(d) illustrates the GND density ρ_{gnd} as a function of the wire radius r. It is interesting to see that the dislocation density in the non-deformed state is pretty high. The density at the beginning decreases when the deformation increases up to γ_{max} = 0.008, whereas the density increases significantly as the plastic strain further increases. Moreover, the position of the highest GND densities evolves from the surface to the center of wires with increasing shear strain. In the strongly plastically deformed wires, GNDs are predominantly concentrated in the center of the wires.

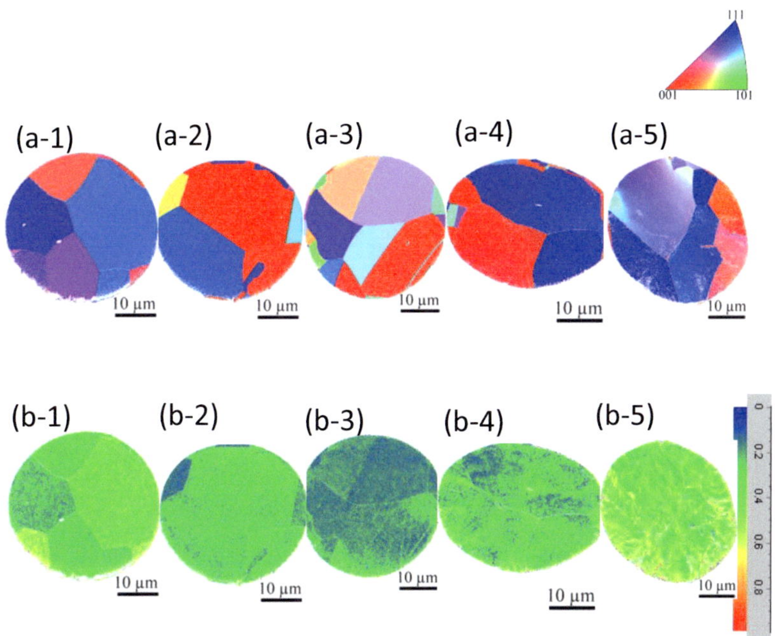

Figure 4.32: Results of EBSD investigations on twisted 40 μm gold wires (annealed at 490°C for 5 hours), loaded at different shear strains of: $\gamma_{max} = 0$, 0.0014, 0.008, 0.08, 0.16 (corresponding to 1-5 in the label, respectively). (a) Orientation maps, (b) misorientation maps

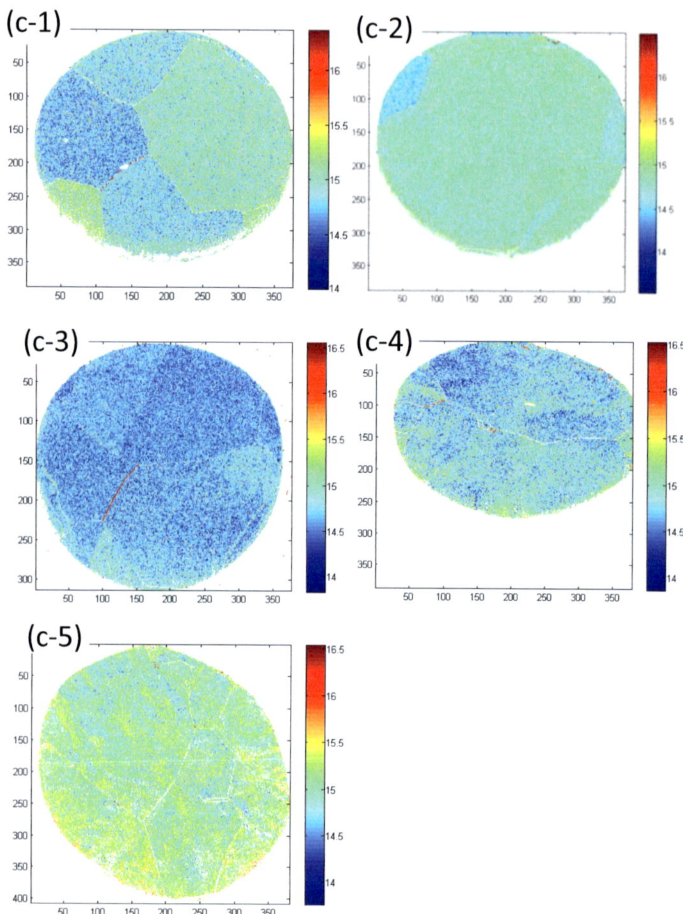

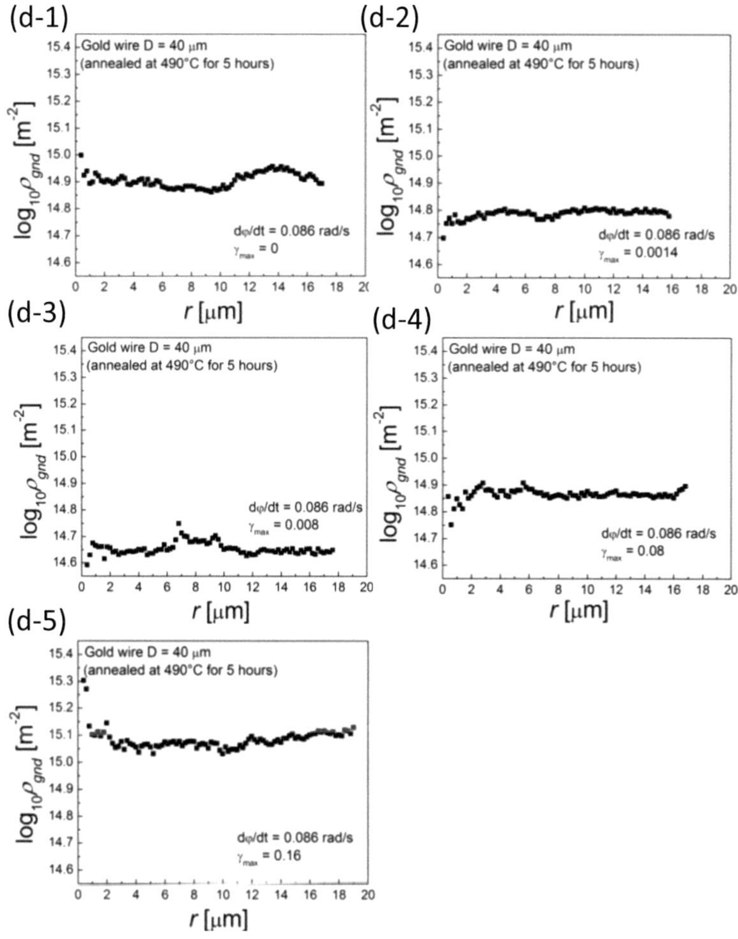

Figure 4.32: Results of EBSD investigations on twisted 40 μm gold wires (annealed at 490°C for 5 hours), loaded at different shear strains of: $\gamma_{max} = 0$, 0.0014, 0.008, 0.08, 0.16 (corresponding to 1-5 in the label, respectively). (c) distribution of GNDs within the cross-sections, (d) distribution of GNDs across the wire radius r. The origin "0" presents the center of the wire.

4.4.2 Dislocation structures in twisted bamboo-structured gold wires

Figure 4.33(a) shows a micrograph of a wire with bamboo structure that was used to investigate the distribution of GNDs in nearly grain boundary-free samples under torsional loading. The different orientations of the single grains along the entire wire length can be clearly observed by color contrast. Figure 4.33(b) shows the micrograph of a cross-section in a bamboo-structured wire. One single crystallite can be seen.

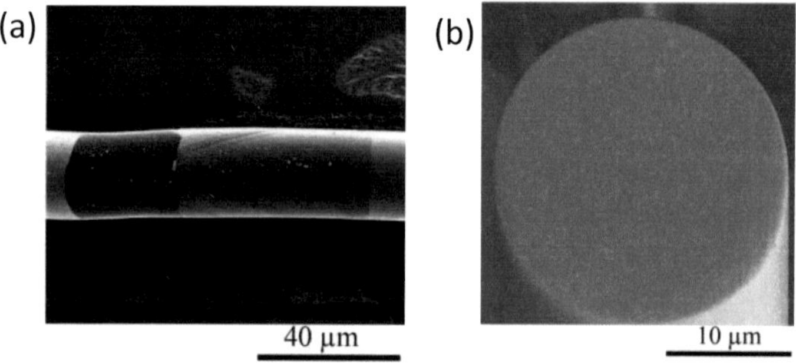

Figure 4.33: Micrographs of a bamboo-structured wire: (a) along the wire length (b) within a cross-section.

Figure 4.34 shows the results of 25 µm gold wires annealed at 740°C for 40 hours loaded up to different shear strains of γ_{max} = 0, 0.006, 0.016, 0.16 (corresponding to 1-4 in Figure 4.34). Figure 4.34(a) shows the orientations of grains within the cross-sections of these wires. The misorientation at the beginning becomes smaller (from bright green to dark green) for small strains, and becomes larger (from bright/dark green to almost yellow) at larger strains. The GNDs in the cross-section mainly concentrate at the surface at small strain levels, as shown in Figure 4.34(c1-c3). However, an inverse-star shape pattern of the distribution of GNDs at large strain is again present, as shown in Figure 4.34(c4). The distribution of GNDs along the wire radius in Figure 4.34(d) confirms the findings from the misorientation maps. The GND density decreases when the shear strain is smaller than γ_{max} = 0.006, and substantially increases with increasing shear strain. Moreover, the concentration of GNDs was found to be close to the center of wires after highest plastic deformation.

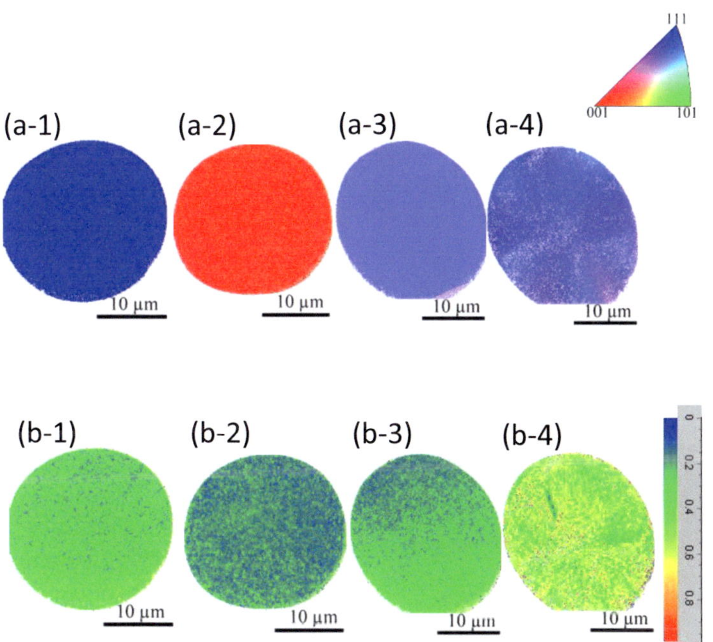

Figure 4.34: Results of EBSD investigations on twisted 25 μm gold wires (annealed at 740°C for 40 hours), loaded at different shear strains of: γ_{max} = 0, 0.006, 0.016, 0.16 (corresponding to 1-4 in the label, respectively). (a) Orientation maps, (b) misorientation maps

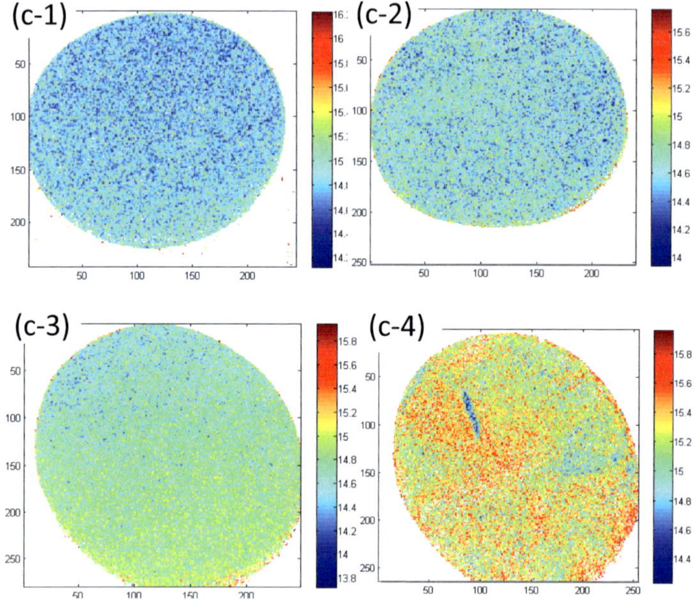

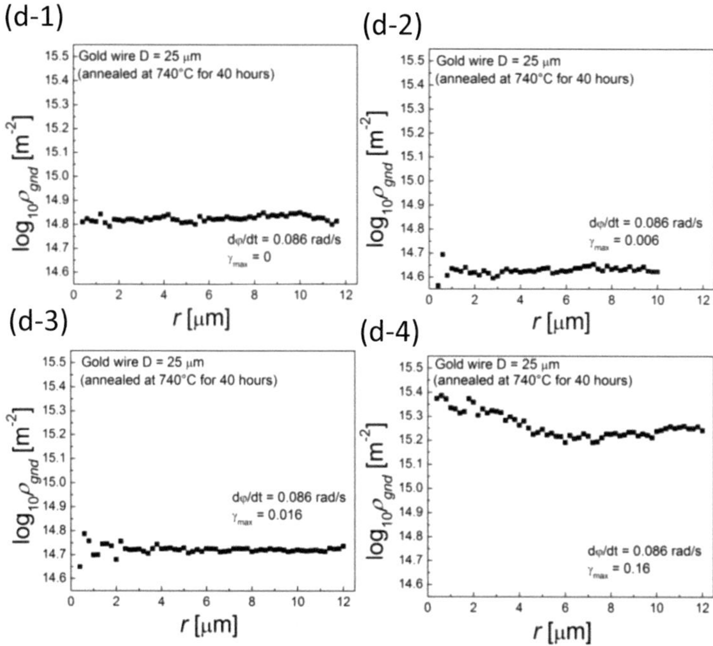

Figure 4.34: Results of EBSD investigations on twisted 25 μm gold wires (annealed at 740°C for 40 hours), loaded at different shear strains of: $\gamma_{max} = 0$, 0.006, 0.016, 0.16 (corresponding to 1-4 in the label, respectively). (c) distribution of GNDs within the cross-sections, (d) distribution of GNDs across the wire radius r. The origin "0" presents the center of the wire.

4.5 Additional investigations on the as-received aluminum wires

In contrast to the noble metal gold or other non-noble fcc metallic materials such as copper, aluminum possesses stable oxide layers at the surface. This feature is useful to compare the mechanical behavior of micro wires in the presence and in the absence of surface oxide layers.

4.5.1 Microstructural characterizations

Figure 4.35 shows cross-sectional micrographs of the aluminum (AlSi1) wires in the as-received state with diameters of 15, 17.5, 25 and 40 µm. The wires were found to have a fine grained structure with average grain sizes d, ranging from 0.33 µm to 0.40 µm. According to the orientation contrast image, especially in Figure 4.35(c) and (d), an annular fabrication texture may be assumed to be present in these wires.

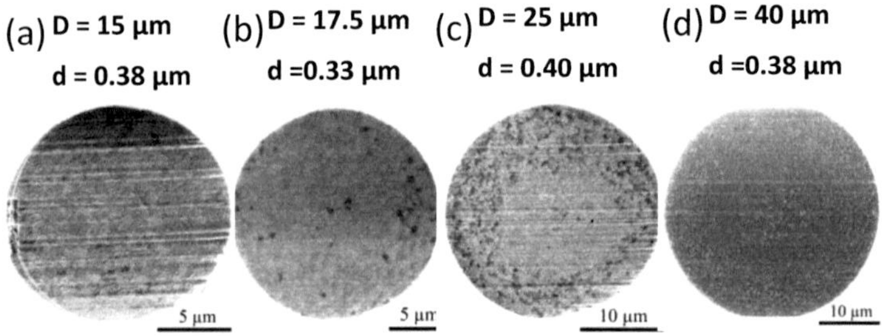

Figure 4.35: FIB micrographs of the cross-sections of as-received aluminum wires with diameter of 15, 17.5, 25, and 40 µm.

Figure 4.36 shows results from Auger electron spectroscopy (AES) investigations on the surface of the aluminum wires. The graphs depict the atomic concentration of Al, O, Si and C in dependence of the sputter depth. The content of O, Si and C was found to be increased in the region close to the surface of the wires. The Si content in the oxide layer is clearly higher (approximately 5 at.% in the 15 and 17.5 µm wire, approximately 9 at.% in the 25 µm wire and 2 at.% in the 40 µm wire) than in the matrix material. Additionally, it was found that the O content in the 15 and 25 µm wires falls down to a low level at a depth of approximately 75 nm, while the O content in the 17.5 and 40 µm wires falls down to a low level at a depth of around 20 nm. Depending on the surface charging, the aluminum content in the oxide layer was not exactly depicted in the graphs. However, according to the change in O content, the oxide layer thickness of 15 and 25 µm wires is approximately 75 nm and the thickness of the layer of the 17.5 and 40 µm wires is approximately 20 nm.

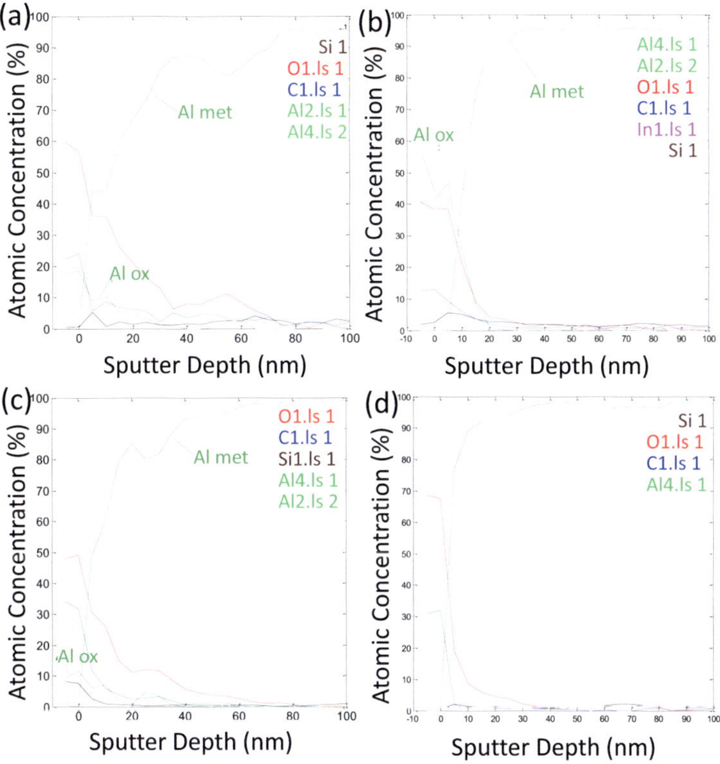

Figure 4.36: Content of Al, O, Si, C in dependence of the sputter depth for the as-received aluminum wires with diameters of: (a) 15 µm, (b) 17.5 µm, (c) 25 µm and (d) 40 µm, determined from Auger electron spectroscopy (AES) investigations.

4.5.2 Deformation behavior in tension and torsion

Figure 4.37 shows the tensile stress-strain behavior of the aluminum wires loaded at different displacement rates. It was found that the strength increases significantly with increasing displacement rate, independent of the wire diameter. A strong hardening was observed after yielding, followed by necking. The uniform elongation ε_u increases slightly at larger displacement rates. However, a systematic trend between the elongation at fracture ε_f and displacement rate was not found.

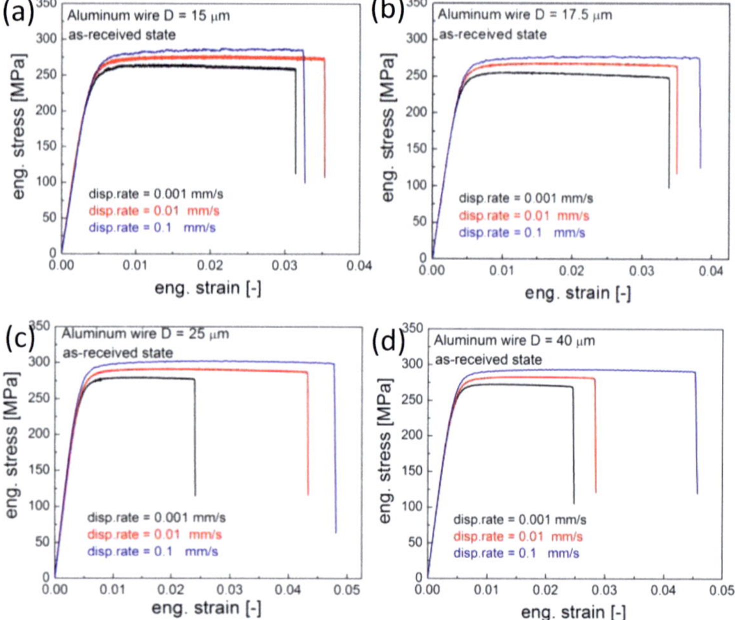

Figure 4.37: The tensile stress-strain behavior of the as-received aluminum wires with diameters of: (a) 15 μm, (b) 17.5 μm, (c) 25 μm and (d) 40 μm as a function of displacement rate.

Figure 4.38 shows the tensile behavior of aluminum wires with various diameters when loaded at different displacement rates. It was found that the 25 µm wire shows the highest ultimate tensile strength, the 40 µm wire shows the second highest one and the 17.5 µm wire has the lowest one ($UTS_{D=25\,\mu m}> UTS_{D=40\,\mu m}> UTS_{D=15\,\mu m}> UTS_{D=17.5\,\mu m}$). This sequence is independent of the displacement rate. When comparing the wires with similar oxide layer thickness, the 40 µm wire shows higher strength than the 17.5 µm wire, and the 25 µm wire exhibits higher strength than the 15 µm wire. A systematic relationship between the elongation at fracture ε_f and the wire diameter was not found.

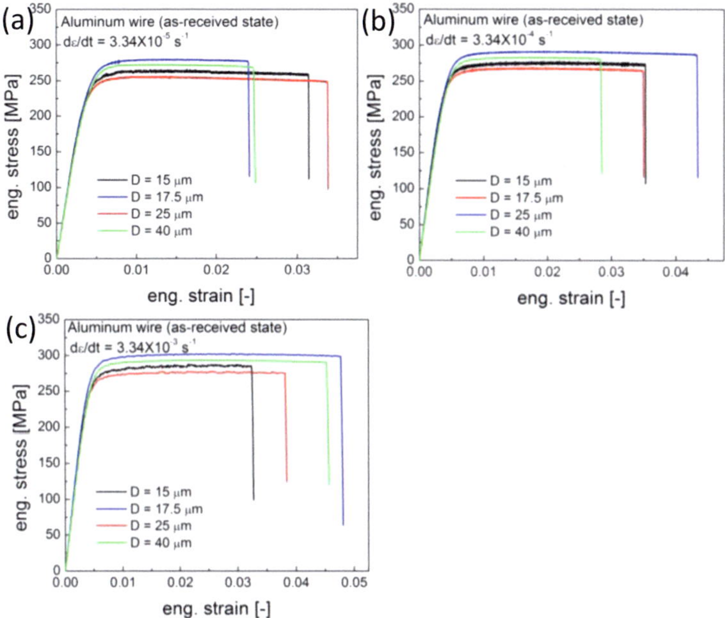

Figure 4.38: The tensile stress-strain behavior of the different thick aluminum wires in the as-received state loaded at displacement rates of: (a) 0.001 mm/s, (b) 0.01 mm/s, and (c) 0.1 mm/s.

Table 4.6 lists the Young's moduli of the aluminum wires determined from tensile unloading segments. The 15 and 25 μm wires exhibit isotropic behavior ($E_{D = 15\ mm} = E_{D = 25\ mm} = 71$ GPa, $E_{iso\text{-}poly} = 70$ GPa [22]), whereas the grains in the 17.5 and 40 μm wires are preferentially oriented in the <100> direction ($E_{D = 17.5\ mm} = E_{D = 40\ mm} = 66$ GPa, $E_{<100>} = 64$ GPa [22]).

Table 4.6: The Young's moduli of the as-received aluminum wires determined from tensile tests with unloading segments

	E0 [GPa]	*E1* [GPa]	*E2* [GPa]	*E3* [GPa]	*E4* [GPa]	*E5* [GPa]	Average value [GPa]
	72	71	71	71	70	70	
15 μm	73	72	72	71	70	71	71
	70	69	69	69	70	69	
	66	66	66	64	65	65	
17.5 μm	67	65	66	66	66	66	66
	65	67	67	66	66	66	
	67	70	71	70	71	70	
25 μm	71	72	71	71	71	71	71
	64	71	71	71	71	71	
	59	65	65	65	65	65	
40 μm	67	65	65	66	66	66	66
	65	66	67	66	67	66	

Under torsional loading, the twist rate has also a significant influence on the deformation behavior of the aluminum wires, as shown in Figure 4.39. It was found that higher twist rates led to higher torsional strength in the aluminum wires, independent of the wire diameter. Additionally, one can observe a slight increase in the hardening with increasing twist rate for larger shear strains.

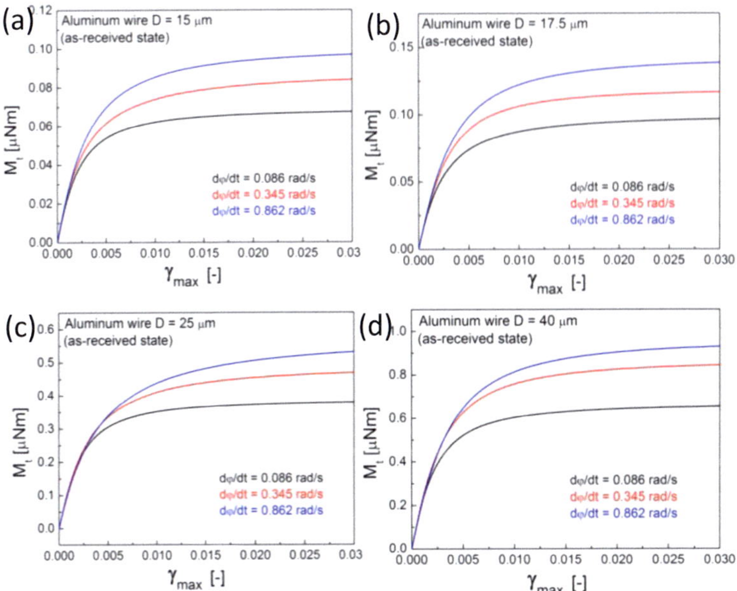

Figure 4.39: Torque behavior of the as-received aluminum wires with diameters of: (a) 15 μm, (b) 17.5 μm, (c) 25 μm, and (d) 40 μm as a function of the angular velocity.

The normalized torsional response of the aluminum wires with various diameters loaded at different angular velocities ($d\varphi/dt$ = 0.086, 0.345 and 0.862 rad/s) is illustrated in Figure 4.40. Similar to the tensile behavior, the 25 μm wires show the highest strength. However, a systematic size effect is evident for 15, 17.5 and 40 μm wires.

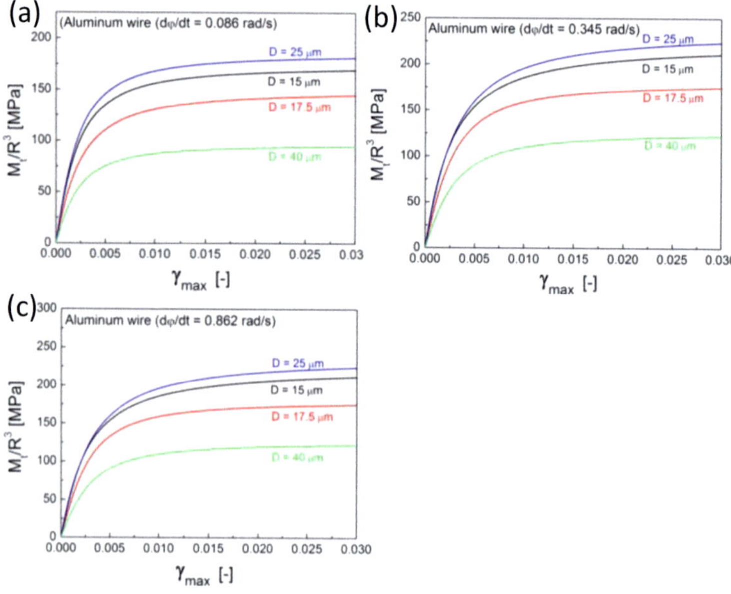

Figure 4.40: Normalized torsional response of the as-received aluminum wires with various diameters loaded at different angular velocities dφ/dt of: (a) 0.086 rad/s, (b) 0.345 rad/s and (c) 0.862 rad/s.

5. Discussion

5.1 Mechanical response of the gold wires

The mechanical investigations generally show that the strain rate does not significantly affect the deformation behavior of the gold wires neither under tensile loading nor under torsional loading. Figure 5.1 shows the 0.2% proof stress ($\sigma_{y0.2}$) and the normalized torque ($M_{t0.2}/R^3$) at 0.2 % γ_{max} as a function of wire diameter (D) for different deformation rates. It can be clearly seen that the wires show no significant strain rate sensitivity. There is only one exceptional case on the as-received 60 µm wires under the torsional loading, as shown in Fig. 5.1(b). Its normalized torque ($M_{t0.2}/R^3$) tends to increase with increasing strain rate, though a normal trend is not observable for the thinner wires in the as-received state. Previous studies on gold wires [146] and gold films [145] found that a strain rate sensitivity was only observed for grain sizes smaller than 500 nm. However, the grain sizes of wires in this study vary between 0.54 µm and 8.7 µm. Thus, it can be seen that the strain rate insensitivity is in good agreement with the previous observations. Moreover, an impact of the specimen size on the normalized torque can be detected for the fully recrystallized wires. The flow stress increases with decreasing wire diameter. This size effect will be discussed later in detail.

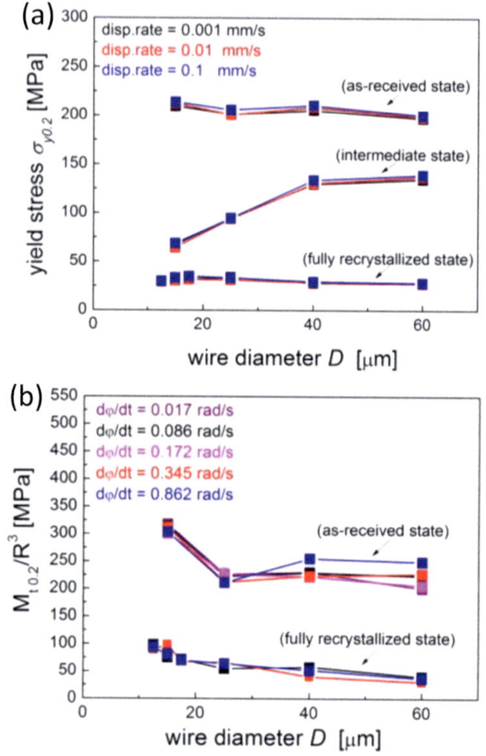

Figure 5.1: Influence of the strain rate on the proof stress for different thick gold wires: (a) in tension, (b) in torsion.

Texture has an impact on both elastic and plastic properties (i.e., yielding and hardening) of a material. Texture was found in the investigated wires due to different degrees of pre-deformation, resulting from cold drawing. The heat treatment investigations (see Figure 4.9) show that the transition temperature from recovery to recrystallization increases with increasing wire diameter. This indicates that the thinner as-received wires exhibit larger degrees of pre-deformation. Besides, the grains at the surface of the as-received wires were stronger deformed during cold drawing than the grains in the center of the wires. Hence, it can be argued that the strongest grains related to the cold working appear at the surface of the thinnest wire and the weakest grains are present in the center of the thickest wire. However, this attribution of grains will be changed during annealing. The grains in the thinnest wire undergo more recovery and recrystallization, which is indicated by the microstructures of the wires in the intermediate state (refer to Figure 4.12). Here, the coarse and presumably weakest grains occur in the region close to the surface of the thinnest wire, while the smallest and presumably strongest grains are present in the center of the thicker wires.

Young's moduli of the gold wires were used in order to identify the texture of the wires. Figure 5.2 illustrates the Young's moduli of the gold wires with diameters of 15, 25, 40 and 60 µm in the three investigated states.

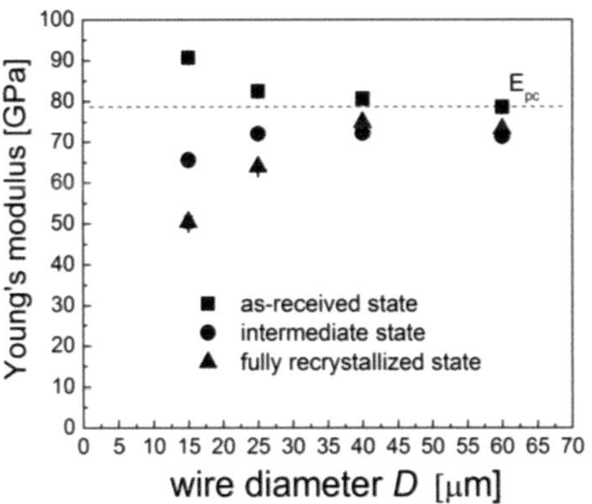

Figure 5.2: Young's moduli of gold wires with diameters of 15, 25, 40 and 60 µm for the three investigated states, determined from unloading segments in tensile tests. The dotted line represents the Young's modulus of isotropic gold.

Compared to the literature values of moduli for bulk gold ($E_{iso\text{-}poly}$ = 78 GPa, $E_{<111>}$ = 117 GPa and $E_{<100>}$ = 43 GPa), it is found that the as-received wires are preferentially oriented in the <111> direction since the determined moduli are higher than the isotropic value. This is more pronounced for the thinner wire. After annealing, the texture changes towards the <100> direction, again more pronounced with decreasing wire diameter.

For fcc metals, the Schmid factor is 0.5 and the maximal resolved shear stress appears at 45° with respect to the tensile axis. When the lattice is oriented in the <100> direction, the maximum resolved shear stress (τ_{max}) lies exactly in the {111} slip planes. This leads to low yield strength (R_e). In contrast, the lattice oriented in the <111> direction results in high yield stress. In agreement with the determined Young's moduli, the yield strength in tension was found to decrease with increasing wire diameter for the as-received wires and increase with increasing wire diameter for the annealed wires (refer to Figure 5.1(a)). Furthermore, the orientation of grains was found to distribute annularly from the surface to the center of the non-recrystallized wires (e.g., in the as-received and in the intermediate states). The grains close to the surface are preferentially oriented in the <100> direction, while the rest of the grains in the wires are dominantly oriented in the <111> direction. Under torsional loadings (σ_n), the onset of plastic deformation starts at the surface, and then evolves towards the center of the wires. In this case, an impact of texture for the torque response can be assumed. For the fully-recrystallized wires, the grains are preferentially oriented in the <100> direction with decreasing wire diameter. Higher stress is required to obtain same resolved shear stress tangent to the circular cross-section for the grains oriented in <100> direction in a warping-free torsion. Consequently, the thinner wire shows the higher stress in torsion and size effect is observed (see Figure 5.1(b)).

5.1.1 Non-fully recrystallized state

As referred in Chapter 4, the uniform elongation ε_u is almost identical to the elongation at fracture ε_f since the specimens used in this study have a very high l_0/D ratio. Both elongations are found to increase with increasing wire diameter in the as-received and intermediate states (see Figure 4.4 and 4.14). Figure 5.3 illustrates the dependency of the uniform elongation ε_u on the wire diameter D at different displacement rates. The increase in uniform elongation with increasing wire diameter is associated with an increase of the critical stress which is necessary to activate necking. Generally, the wires exhibit an almost perfect cylindrical shape with smooth surfaces. Failure is observed to occur at random positions along the length of the wire. One may assume that necking is triggered by small surface defects. It is conceivable that the critical load to activate necking process decreases with decreasing wire diameter when the size of such defects is independent of the wire diameter.

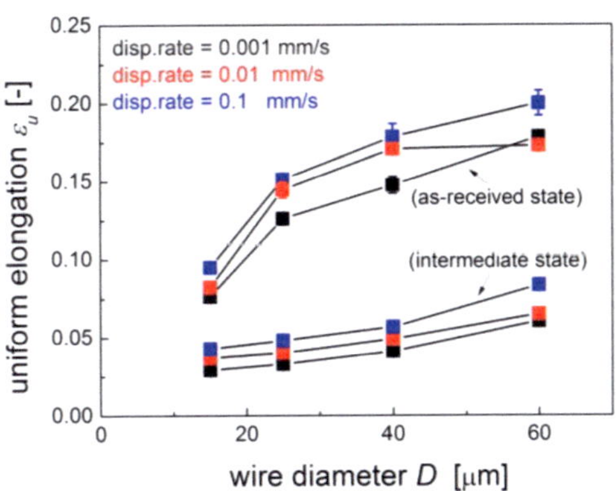

Figure 5.3: Uniform elongation ε_u for the non-fully recrystallized gold wires loaded at different displacement rates as a function of wire diameter D.

Furthermore, in tension, the as-received wires show almost ideal elastic-plastic behavior, whereas the thinner wires show stronger hardening in the intermediate state. This may be related to the texture difference of these cold drawn wires. In torsion, the as-received wires show almost ideal plastic behavior at large strains. The hardening exponent n for all wire diameters and strain rates is equal ($n = 31\pm1$) at large plastic deformation (The determination of n refers to Ramberg-Osgood fit in Section 3.4.2.3). The high value of hardening exponent n is associated with the high dislocations retained from the pre-deformation during the cold drawing. They are easier to meet each other and lead to strong hardening at low strains. In the intermediate state, the 25 and 60 µm wires show similar behavior at large strains. In contrast, the 15 µm ($n = 4$) and 38 µm ($n = 4.2$) wires show continuous hardening with increasing shear strain. Hence, no consistent trend between the hardening and wire diameter was found. This can be traced back to the inhomogeneous microstructures with different degrees of recrystallization.

In the non-fully recrystallized states, the size depending inhomogeneous microstructures with varying local grain sizes and dislocation densities result in a strong superimposed Hall-Petch effect. Figure 5.4 shows the 0.2% proof stress ($\sigma_{y0.2}$) and normalized torque ($M_{t0.2}/R^3$) at 0.2 % γ_{max} of the wires as a function of the grain size ($d^{1/2}$). It was found that the proof stress in the non-fully recrystallized state can be fitted with the Hall-Petch relation but this results in negative values of the friction stress σ_o. The influence of cold work hardening was found to be superimposed on the Hall-Petch effect for the non-fully recrystallized wires. Accordingly, it is impossible to verify the influence of strain gradients on the size effect besides the Hall-Petch effect from the non-fully recrystallized wires, due to the superimposed impact from the cold work.

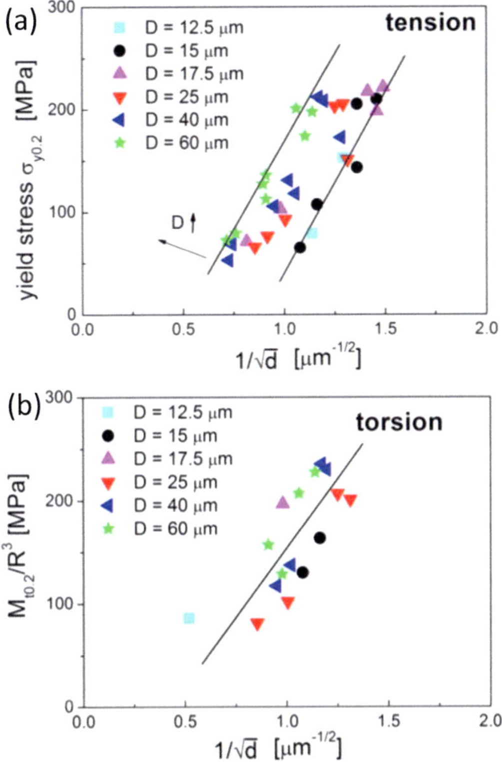

Figure 5.4: Hall-Petch behavior of the gold wires in the non-fully recrystallized state: (a) in tension, (b) in torsion.

5.1.2 Fully recrystallized state

In the fully recrystallized state, a general trend that the uniform elongation increases with increasing wire diameter is found. However, a systematic relationship between the uniform elongation and the wire diameter is no longer observed, as one can see from Figure 5.5. One may argue that surface defects and the local orientation of the grains at the surface have a concurrent effect. For the coarse-grained structures, the local orientation of the grains at the surface has a significant impact on the onset of necking process. The defects on the grains, which are well-oriented in preferential directions such as <100> direction, may possibly induce necking processes at lower load.

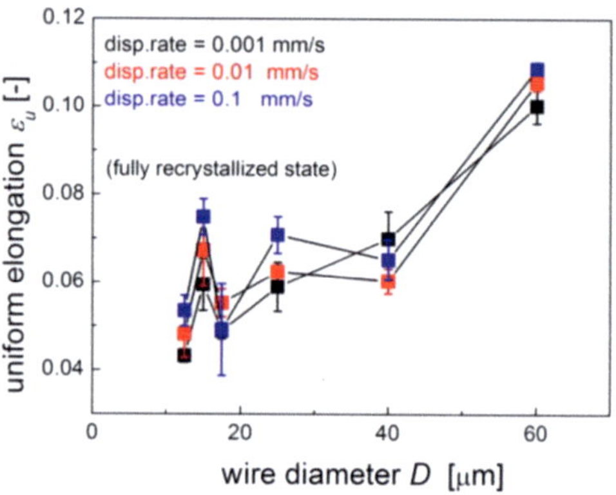

Figure 5.5: Uniform elongation ε_u of the fully recrystallized gold wires at different displacement rates as a function of the wire diameter D.

Figure 5.6 illustrates the 0.2% proof stress $\sigma_{y0.2}$ (Figure 5.6(a)) and the normalized torque $M_{t0.2}/R^3$ at 0.2 % γ_{max} (Figure 5.6(b)) in dependence of grain size $d^{1/2}$ in tension and torsion. Hall-Petch effect can be evidently observed in both tension and torsion. Nevertheless, the Hall-Petch behavior of gold wires in tension was found to be size dependent, as shown in Figure 5.6(a). In tension, the thinner wires oriented more in the <100> direction is easier to be deformed. Simultaneously, depend on the surface weakening effect, a higher proportion of dislocations in the thinner wires may disappear on the free surface and result in the lower 0.2% proof stress ($\sigma_{y0.2}$) of the wires. In contrast, no clear size dependency is seen for the data in torsion for the different thick wires, as shown in Figure 5.6(b), where the data points fall more or less onto one curve. In torsion, as mentioned above, the texture difference may lead to higher stress in the thinner wire. However, the impact of free surface on the dislocations may weak the strength of the thinner wires. The counterbalance of the concurrent effects may be associated with the size independency of the Hall-Petch behavior for the fully-recrystallized wires in torsion.

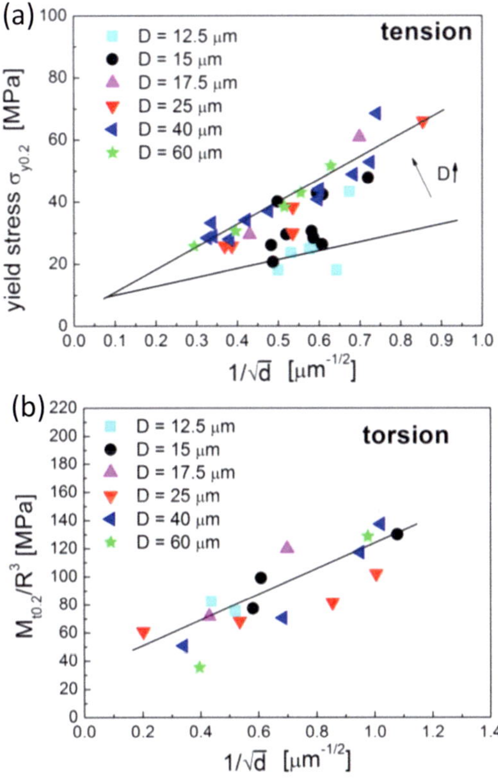

Figure 5.6: Hall-Petch behavior of the fully recrystallized gold wires: (a) in tension, (b) in torsion.

Previous studies [135, 166, 167] have shown that in tension the strength deviates from the Hall-Petch behavior for small enough specimen to grain size ratios D/d. For the coarse grained structure, dislocations may leave easily the grains at the free surface during tensile deformation since there are not so much grain boundaries to constrain the movement of dislocations. Therefore, the ratio (D/d) between grain size d and the wire diameter D becomes important. However, as we can see from the microstructure in Figure 4.16, the 15, 25 and 40 μm wires have similar numbers of grains across the cross-sections for the different thick wires, indicating that the wires have similar ratios D/d. However, their behavior is different and it is argued that D/d may not be the appropriate parameter for describing the strength of the investigated wires as a function of grain and specimen size.

Figure 5.7(a) shows that the hardening exponent n in tension increases with decreasing wire diameter. It is observed that the thicker wires have lower hardening exponents (n) which indicates stronger hardening. This is consistent with the tensile stress-strain behavior of the fully recrystallized wires in Figure 4.19. Figure 5.7(b) shows the normalized hardening rate of the plastic deformation of these gold wires in tension. The values were derived by mean of smoothing the original determined true stress-strain curve. It was found that the normalized hardening rate decreases with increasing strain. Moreover, the thicker wires show a stronger decrease rate related to higher hardening.

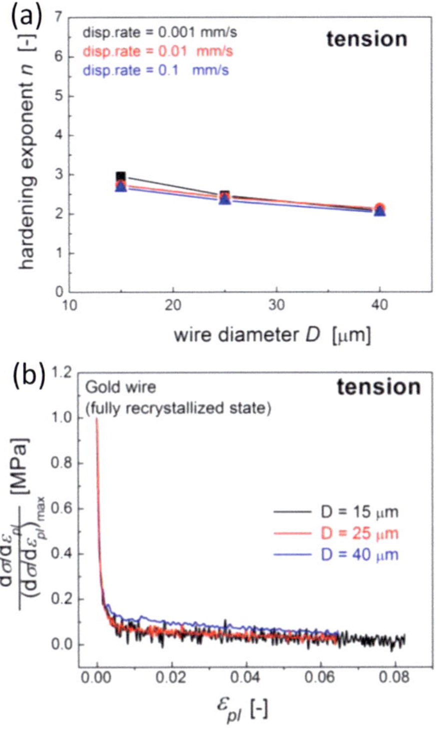

Figure 5.7: Hardening behavior of the fully recrystallized gold wires with diameters of 15, 25 and 40 µm in tension. (a) Hardening exponent n as a function of the wire diameter D at different displacement rates, and (b) normalized hardening rate as a function of the plastic strain.

On one hand, as mentioned above, the wires show similar numbers of grains across the cross-sections. Thereby, the thicker wires exhibit a higher surface- and grain boundary-to-volume ratio. After the initiation of plastic deformation, the probability for dislocations to escape the free surface is lower in thicker wires. In other words, the dislocations are easier to meet and tangle with each other in the grains, which lead to stronger mechanical hardening for the thicker wires. On the other hand, the larger grain boundary length in thicker wires offers more dislocation sources for the generation of new dislocations. The interactions of these new generated dislocations also contribute to the stronger hardening in thick wires.

Figure 5.8(a) illustrates the hardening exponent n in torsion as a function of the wire diameter D. A slight increase in hardening with decreasing wire diameter is observed. However, the effect is quite weak and almost negligible. Figure 5.8(b) shows the normalized hardening rate of the plastic deformation of these gold wires in torsion. The curves are derived from the Ramberg-Osgood fitted curves. It was found that the normalized torque decreased with increasing shear strain at the surface. The decrease rate of the normalized torque is larger in the thinner wire. Thus, the thinner wire shows stronger hardening at low shear strains ($\gamma_{max} < 0.03$) (see Figure 4.21). Under torsional loadings, GNDs are introduced based on the strain gradients. For the same shear strain at surface (γ_{max}), the thinner wires have larger strain gradients. More GNDs are generated in thinner wire to accommodate the lattice curvature. These GNDs meet the initial dislocations and piles up, which contributes to the higher hardening. Besides, the thinner wire shows smaller grain size. The dislocations may meet the grain boundaries easier under the shear stress in the thinner wires and result in the slightly stronger hardening.

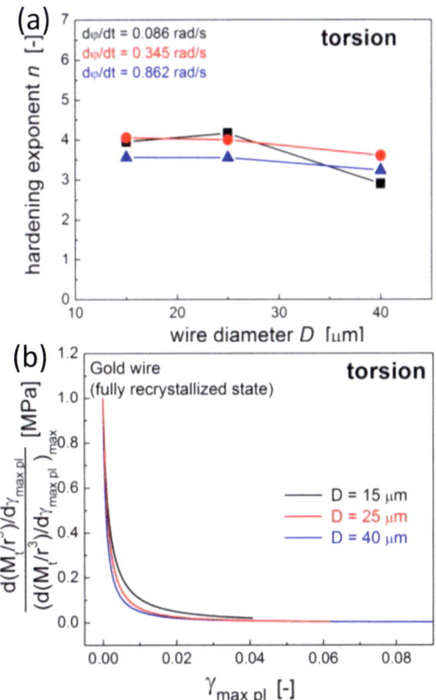

Figure 5.8: Hardening behavior of gold wires with diameter of 15, 25 and 40 µm in the fully recrystallized state in torsion. (a) Hardening exponent n as a function of the wire diameter D for different angular rates, and (b) normalized hardening rate as a function of the plastic shear strain.

Grain boundaries play a significant role in the hardening behavior of wires. The impact of grain boundaries can be seen more clearly when comparing the deformation behavior of polycrystalline and bamboo-structured wires. Figure 5.9 shows the comparison of the torsional response of a 25 µm coarse grained wire (annealed at 450°C for 5 hours) and a 25 µm bamboo-structure wire (annealed at 740°C for 40 hours). The polycrystalline wire shows significant stronger hardening than the bamboo structured wire.

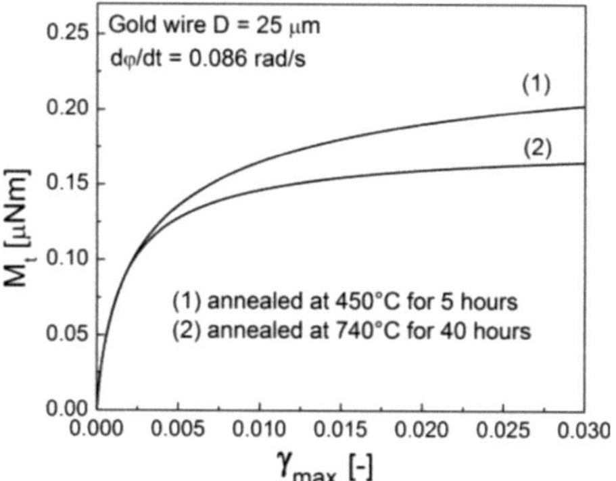

Figure 5.9: Torsional response of the 25 μm gold wires with both coarse-grained structure (annealed at 450°C for 5 hours) and bamboo structure (annealed at 740°C for 40 hours).

A size effect is observed in torsion for all the investigated states. The size effects under torsional loadings are found to result from a combination of different degrees of pre-deformation (generated during the fabrication process of the different thick wires by cold drawing), different degrees of recrystallization, the Hall-Petch effect, texture differences and the occurrence of strain gradients. In the fully recrystallized state, the influence of initial dislocation density due to the pre-deformation from cold drawing process is no longer present. As mentioned in Section 5.1, the grains are more oriented in the <100> direction in thinner wires. The texture difference of different thick wires is assumed to have an impact on the observed size effect. Besides, the Hall-Petch effect is confirmed on the fully-recrystallized 15, 25 and 40 μm since the grain size (d) of these wires is increasing with increasing wire diameter (D). Therefore, it can be concluded that strain gradients are not responsible for the size effect when the grain size scales with the wire diameter. Moreover, as it is not possible to have the same grain size when comparable deformation behavior in tension is required on these wires, it is difficult to separate the superimposed effect of Hall-Petch effect from the gradient effect.

Single-crystal wires would be a good option if they could be fabricated. Other methods can be also considered to obtain the wires with different diameters from one original wire by cutting using FIB or reducing the diameter by etching. In any case, further investigations on wires with comparable grain size in the fully recrystallized state are needed to separate the influence of grain size. This will provide a possibility to distinguish the influence of strain gradients from the influence of the microstructure on the size effect.

5.2   The distribution of GNDs

The distribution of GNDs was systematically investigated on polycrystalline and bamboo-structured gold wires for different shear strains. For low strains (see Fig. 4.31(c1-c2), 4.32(c1-c3) and 4.34(c1-c3)), no clear structures (mis-/orientation, distribution of GNDs) or structural changes are observed. The method is obviously not sensitive enough to get clear results concerning dislocation structures. The deformation behavior of wires is strongly affected by local microstructural boundary conditions related to the individual cross-sections studied. For finer grained structures (see Fig. 4.31(c1-c2)), the influence of grain boundaries on the distribution of GNDs seems to be dominant compared to the influence of gradients. For coarse grained (see Fig. 4.32(c1-c3)) and bamboo structures (see Fig. 4.34(c1-c3)), the results reveal an inhomogeneous deformation along the entire length of these wires. Therefore, in order to get reliable information about the progress of GND densities, it is very important to cut wires at positions where the textural conditions are similar for different degrees of deformation. Furthermore, the highest GND density was found close to the surface of wires for low strains as the plastic deformation in torsion starts at the surface of wires.

For large shear strains (see Fig. 4.31(c3), 4.32(c4-c5) and 4.34(c4)), the distribution of GNDs shows an inverse-star shape pattern within the investigated cross-sections. This might be related to two general conditions. One is that initial dislocations can lead to a preferred localized deformation with preferred deformation paths, causing lines from the surface towards the center. The other one is due to the principle of energy minimization. A high density of dislocations implies a high inner energy. Therefore, dislocation structures might be formed such as small angle grain boundaries to lower this energy. These sub-boundaries may evolve to the inverse-star shape pattern.

The highest GND density is found evolving from the surface to the center in polycrystalline gold wires with increasing shear strain (refer to Figure 4.31(d) and 4.32(d)). However, the highest GND density is in the center of the wire when the shear strain is 0.006 (refer to Figure 4.34(d)), indicating that the GNDs in a bamboo-structured wire seem easier to be driven towards the center without the restriction of grain boundaries. Since the neutral fiber is at the center for a wire under torsion, there is a stress gradient between the surface and the center of wires. Besides, there is always high GND density in the center of strongly deformed wires, even for polycrystalline wires. This results in an additional stress field. As a result, dislocations will be driven to the center of wires, where the stress is low.

5.3 Mechanical response of as-received aluminum wires

Strain rate sensitivity is found in the strength of the aluminum wires with diameters of 15, 17.5, 25 and 40 μm under both tensile and torsional loadings. Figure 5.10 plots the 0.2% proof stress ($\sigma_{y0.2}$) and torque ($M_{t0.2}$) at 0.2 % γ_{max} as a function of wire diameter (D) for different strain rates. Higher strength levels were observed at larger strain rates. The strain-rate sensitivity was found to depend strongly on the grain size [147], where there is critical grain size to separate the rate-sensitive and rate-insensitive regime. In this study, the mean grain size of the aluminum wires is about 0.35 µm, which may fall into the rate-sensitive regime for aluminum. Additionally, the oxide layer may also have an impact on the strain rate sensitivity of aluminum wires. Based on the oxide layer at the surface, the grains, especially the ones close to the surface region, are constrained and difficult to be deformed. Since the plastic deformation starts at the surface, an impact of the oxide layer under torsional loading can be expected.

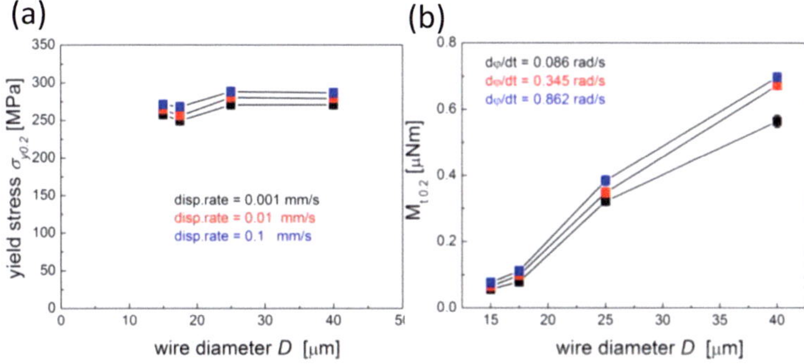

Figure 5.10: Influence of strain rate on the 0.2 % proof stress and torque for aluminum wires with diameters of 15, 17.5, 25 and 40 µm: (a) in tension, (b) in torsion.

The aluminum wires exhibit significant length extension after the onset of necking. In contrast to gold, the uniform elongation ε_u and the elongation at fracture ε_f of aluminum wires are not identical any more. Figure 5.11 illustrates the uniform elongation (ε_u) and the elongation at fracture (ε_f) of aluminum wires at different strain rates as a function of wire diameter (D). There is no systematic relationship found between the uniform elongation (ε_u)/elongation at fracture (ε_f) and the wire diameter. However, it can be noticed that the uniform elongation (ε_u) of the wires increases with increasing displacement rate.

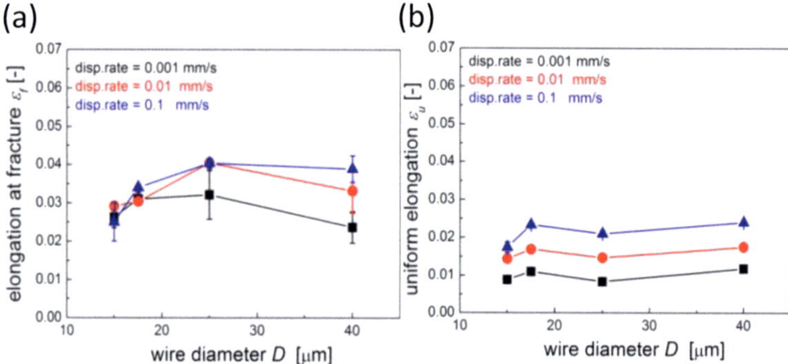

Figure 5.11: Elongation of the as-received aluminium wires as a function of wire diameter: (a) uniform elongation ε_u, (b) elongation at fracture ε_f.

A specimen size effect in the flow stress is observed for the aluminum wires in torsion which is not present in tension, although the 15 μm wires exhibit a slightly higher strength in tension compared to the 17.5 μm wires. This most likely related to the different degrees of pre-deformation of the wires with a highly strain-hardened surface region. Besides, the thickness of the oxide layer may also have stronger impact on the torque behavior of the wires than on the tensile behavior.

6.   Conclusions and outlook

The deformation behavior of different thick gold micro wires with varying microstructures was investigated under both tensile and torsional loadings. There is no strain rate dependency observed on the proof stress of the investigated wires neither in tension nor in torsion. This may be associated with the grain size of these wires.

For all wires, the relationship between grain size and the annealing temperature / dwell time was obtained during systematic heat treatment investigations. Consequently, it was possible to create microstructures in different states, i.e., non-recrystallized state or fully-recrystallized state.

For the non-fully recrystallized wires, the grains close to the surface and the center of the wires are preferentially oriented in the <100> direction, despite the overall dominant orientation of the wires is in the <111> direction. For the fully-recrystallized wires, the main orientation is evolving towards the <100> direction without distinct distribution of grains from the surface to the center. The grains are preferentially oriented towards the <111> direction with decreasing wire diameter in the as-received wires, whereas the grains are preferentially oriented towards the <100> direction with decreasing wire diameter in the annealed wires. This is associated with the different degrees of recovery / recrystallization of the annealed wires due to the previous cold drawing processes. An impact of the texture on both elastic and plastic properties (i.e., yielding and hardening) was found.

Hall-Petch behavior was observed in both tension and torsion for the fully-recrystallized wires. The Hall-Petch effect in tension shows a specimen size dependency, while it is size independent in torsion. This difference is argued to be related to the directionality of dislocation movements for the different loading types. It is argued that dislocations may leave the grains at the free surface during tensile deformation. However, the geometrically necessary dislocations are driven towards the center of wires in torsion depending on the additional stress-strain field.

Stronger hardening was found on the thicker wires in tension for the 15, 25, 40 μm wires, which have approximately the same ratio D/d. For equal dislocation densities, it is assumed that the probability for dislocations to escape from the free surface is lower in the thicker wires. In this case, the higher absolute number of dislocations in thicker wires leads to a higher probability of dislocations to interact with each other before they reach free surfaces. Moreover, the larger grain boundary length in thicker wires offers more dislocation sources for the generation of new dislocations. These new generated dislocations move and pile up again, which also contributes to the high hardening in thick wires.

However, it is difficult to separate the superimposed effect of Hall-Petch from the strain gradient effects. Further investigations on the mechanical properties of fully-recrystallized wires with comparable grain size may be helpful to distinguish the influence of grain sizes and the influence of strain gradients on the size effect.

Furthermore, the distribution of GNDs was found to be affected by strain gradients and grain boundaries. The grain boundaries play a dominant role at small strains in fine grained structures. In contrast, the influence of strain gradients becomes evident at large strains in coarse-grained structures. The GNDs seem to be driven towards the center of bamboo-structured wires without the obstacle of grain boundaries. With increasing shear strain, the highest GND densities were found evolving from the surface to the center of the wires. This is assumed to be associated to the stress field between the surface and the center of wires. Besides, it seems that the GND density decreases first and then increases with increasing shear strain, which is strongly affected by local microstructural boundary conditions within individual observed cross-sections. Therefore, in order to get reliable information about the progress of GND densities, it is important to cut the wires at positions where the textural conditions are similar.

In contrast to gold wires, strain rate sensitivity was found on the proof stress of aluminum wires in both tension and torsion. The uniform elongation ε_u of aluminum wires was found to increase with increasing displacement rate. The 17.5 and 40 μm wires have similar thickness of oxide layer around 20 nm, while the 15 and 25 μm wires have similar thickness of oxide layer around 75 nm. Size effect is observed for the aluminum wires with diameters of 15, 17.5 and 40 μm in torsion. However, the 25 μm wire still shows higher proof stress than the 15 μm wire though they have similar thick oxide layer. This may be related to the different degrees of pre-deformation of the wires due to the cold drawing processes. Besides, the thickness of an oxide layer may also have an impact on the torque behavior of the wires. Consequently, fully recrystallized different thick aluminum wires with comparable microstructures should be fabricated and tested in further investigations. A comparison with the present results from the gold wires may lead to the possibility to clarify the underlying mechanisms on materials at micron scales.

Based on the systematic experimental results, the influence of different parameters on the deformation behavior may be quantitatively described or simulated. This makes it possible to predict the mechanical properties of materials in small dimensions.

References

[1] M. Antler. Gold in electrical contacts. *Gold Bulletin,* 4(3): 42-46, 1971.

[2] N.A. Fleck, J.W. Hutchinson. A phenomenological theory for strain gradient effects in plasticity. *Journal of the Mechanics and Physics of Solids*, 41(12): 1825-1857, 1993.

[3] N.A.Fleck, J.W. Hutchinson. Strain Gradient Plasticit. in *Advances in Applied Mechanics*, Elsevier, 295-36, 1997.

[4] E.C. Aifantis. Strain gradient interpretation of size effects. *International Journal of Fracture*, 95(1-4): 299-314, 1999.

[5] S.H. Chen, T.C. Wang. A new hardening law for strain gradient plasticity. *Acta Materialia*, 48(16): 3997-4005, 2000.

[6] S.H. Chen, T.C. Wang. A new deformation theory with strain gradient effects. *International Journal of Plasticity*, 18(8): 971-995, 2002.

[7] H. Gao, Y. Huang. Taylor-based nonlocal theory of plasticity. *International Journal of Solids and Structures*, 38(15): 2615-2637, 2001.

[8] H. Gao, Y. Huang, W.D. Nix, J.W. Hutchinson. Mechanism-based strain gradient plasticity-- I. Theory. *Journal of the Mechanics and Physics of Solids,* 47(6): 1239-1263, 1999.

[9] Y. Huang, H. Gao, W.D. Nix, J.W. Hutchinson. Mechanism-based strain gradient plasticity--II. Analysis. *Journal of the Mechanics and Physics of Solids*, 48(1): 99-128, 2000.

[10] X. Qiu, Y. Huang, Y. Wei, H. Gao, K.C. Hwang. The flow theory of mechanism-based strain gradient plasticity. *Mechanics of Materials*, 35(3-6): 245-258, 2003.

[11] Y. Huang, S. Qu, K.C. Hwang, M. Li, H. Gao. A conventional theory of mechanism-based strain gradient plasticity. *International Journal of Plasticity*, 20(4-5): 753-782, 2004.

[12] S. Brinckmann, T. Siegmund, Y. Huang. A dislocation density based strain gradient model. *International Journal of Plasticity*, 22(9): 1784-1797, 2006.

[13] A. Acharya, J.L. Bassani. Lattice incompatibility and a gradient theory of crystal plasticity. *Journal of the Mechanics and Physics of Solids*, 48(8): 1565-1595, 2000.

[14] M.E. Gurtin. A gradient theory of single-crystal viscoplasticity that accounts for geometrically necessary dislocations. *Journal of the Mechanics and Physics of Solids*, **50**(1): 5-32, 2002.

[15] A.J. Bushby, D.J. Dunstan. Size effects in yield and plasticity under uniaxial and non-uniform loading: experiment and theory. *Philosophical Magazine*, 91(7-9): 1037-1049, 2011.

[16] N.A. Fleck, G.M. Muller, M.F. Ashby, J.W. Hutchinson. Strain Gradient Plasticity - Theory and Experiment. *Acta Metallurgica Et Materialia,* 42(2): 475-487, 1994.

[17] D. J. Dunstan, B. Ehrler, R. Bossis, S. Joly, K. M.Y. P'ng, and A. J. Bushby. Elastic Limit and Strain Hardening of Thin Wires in Torsion. *Physical Review Letters*, 103(15): 155501, 2009.

[18] D. Liu, Y. He, X. Tang, H. Ding, P. Hu, P. Cao. Size Effects in the Torsion of Microscale Copper Wires: Experiment and Analysis. *Scripta Materialia*, 66(6): 406-409, 2012.

[19] J.F. Nye. Some geometrical relations in dislocated crystals. *Acta Metallurgica*, 1(2): 153-162, 1953.

[20] M.F. Ashby. The deformation of plastically non-homogeneous materials. *Philosophical Magazine*, 21(170): 399-424, 1970.

[21] T.T. Zhu, A.J. Bushby, D.J. Dunstan. Materials mechanical size effects: a review. *Materials Technology*, 23(4): 193-209, 2008.

[22] T.H. Courtney. Mechanical Behavior of Materials. New York: McGraw-Hill., 1990. ISBN:0070132658.

[23] C.S. Han, H. Gao, Y. Huang, W.D. Nix. Mechanism-based strain gradient crystal plasticity--II. Analysis. *Journal of the Mechanics and Physics of Solids*, 53(5): 1204-1222, 2005.

[24] D.M. Duan, N.Q. Wu, W.S. Slaughter, S.X. Mao. Length scale effect on mechanical behavior due to strain gradient plasticity. *Materials Science and Engineering A*, 303(1-2): 241-249, 2001.

[25] A. Arsenlis, D.M. Parks. Crystallographic aspects of geometrically-necessary and statistically-stored dislocation density. *Acta Materialia*, 47(5): 1597-1611, 1999.

[26] C.S. Busso, K.M. Devos, G. Ross, M. Mortimore, W.M. Adams, M.J. Ambrosel, S. Alldrick, M.D. Gale. Genetic diversity within and among landraces of pearl millet (Pennisetum glaucum) under farmer management in West Africa. *Genetic Resources and Crop Evolution,* 47(5): 561-568, 2000.

[27] A. Acharya, A.J. Beaudoin. Grain-size effect in viscoplastic polycrystals at moderate strains. *Journal of the Mechanics and Physics of Solids*, 48(10): 2213-2230, 2000.

[28] J.L. Bassani. Incompatibility and a simple gradient theory of plasticity. *Journal of the Mechanics and Physics of Solids*, 49(9): 1983-1996, 2001.

[29] A.J. Beaudoin, A. Acharya. A model for rate-dependent flow of metal polycrystals based on the slip plane lattice incompatibility. *Materials Science and Engineering A*, 309-310(0): 411-415, 2001.

[30] L.P. Eversa, D.M. Parks, W.A.M. Brekelmans, M.G.D. Geers. Crystal plasticity model with enhanced hardening by geometrically necessary dislocation accumulation. *Journal of the Mechanics and Physics of Solids*, 50(11): 2403-2424, 2002.

[31] R. Deborst, H.B. Muhlhaus, Gradient-Dependent Plasticity - Formulation and Algorithmic Aspects. *International Journal for Numerical Methods in Engineering*, 35(3): 521-539, 1992.

[32] N.A. Fleck, J.W. Hutchinson. A reformulation of strain gradient plasticity. *Journal of the Mechanics and Physics of Solids*, 49(10): 2245-2271, 2001.

[33] H. Gao, Y. Huang, W.D. Nix. Modeling plasticity at the micrometer scale. *Naturwissenschaften*, 86(11): 507-515, 1999.

[34] Y. Huang, Z. Xue, H. Gao, W.D. Nix, Z.C. Xia. A study of microindentation hardness tests by mechanism-based strain gradient plasticity. *Journal of Materials Research*, 15(8): 1786-1796. 2000.

[35] X. Qiu, Y. Huang, W.D. Nix, K.C. Hwang, H. Gao. Effect of intrinsic lattice resistance in strain gradient plasticity. *Acta Materialia*, 49(19): 3949-3958, 2001.

[36] M.E. Gurtin. On the plasticity of single crystals: free energy, microforces, plastic-strain gradients. *Journal of the Mechanics and Physics of Solids*, 48(5): 989-1036, 2000.

[37] K.C. Hwang, H. Jiang, Y. Huang, H. Gao. A finite deformation theory of strain gradient plasticity. *Journal of the Mechanics and Physics of Solids*, 50(1): 81-99, 2002.

[38] K.C. Hwang, H. Jiang, Y. Huang, H. Gao. Finite deformation analysis of mechanism-based strain gradient plasticity: torsion and crack tip field. *International Journal of Plasticity*, 19(2): 235-251, 2003.

[39] P.M. Mariano. A note on Ceradini-Capurso-Maier's theorem in plasticity. International Journal of Plasticity, 18(12): 1749-1773, 2002.

[40] W. Wang, Y. Huang, K.J. Hsia, K.X. Hu, A. Chandra. A study of microbend test by strain gradient plasticity. *International Journal of Plasticity*, 19(3): 365-382, 2003.

[41] E.C. Aifantis. On the microstructural origin of certain inelastic models. *Journal of Engineering Materials and Technology. Transactions of the American Society of Mechanical Engineers*, 106: 326-330, 1984.

[42] E.C. Aifantis. The physics of plastic deformation. *International Journal of Plasticity*, 3: 211-247, 1987.

[43] E.C. Aifantis. On the role of gradients in the localization of deformation and fracture. *International Journal of Engineering Science*, 30: 1279-1299, 1992.

[44] H.B. Mühlhaus, E.C. Aifantis. A variational principle for gradient plasticity. *International Journal of Solids and Structures*, 28: 845-857, 1991.

[45] J.L.M. Morrison. The yield of mild steel with particular reference to the effect of size of specimen. *Proceedings of the Institute of Mechanical Engineers*, 142: 193-223, 1939.

[46] H.T. Zhu, H.M. Zbib. Strain gradients and continuum modeling of size effect in metal matrix composites. *Acta Mechanica*, 121(1-4): 165-176, 1997.

[47] C.W. Richards. Effect of size on the yielding of mild steel beams.

Proceedings of the American Society for Testing and Materials, 58: 955-970, 1958.

[48] J.S. Stölken, A.G. Evans. A microbend test method for measuring the plasticity length scale. *Acta Materialia*, 46(14): 5109-5115, 1998.

[49] Z.C. Xia, J.W. Hutchinson. Crack tip fields in strain gradient plasticity. *Journal of the Mechanics and Physics of Solids*, 44(10): 1621-1648,1996.

[50] Y. Huang, L. Zhang, T.F. Guo, K.C. Hwang. Mixed mode near-tip fields for cracks in materials with strain-gradient effects. *Journal of the Mechanics and Physics of Solids*, 45(3): 439-465, 1997.

[51] M.R. Begley, J.W. Hutchinson. The mechanics of size-dependent indentation. *Journal of the Mechanics and Physics of Solids*, 46(10): 2049-2068, 1998.

[52] J.Y. Shu, N.A. Fleck. Strain gradient crystal plasticity: size-dependent deformation of bicrystals. *Journal of the Mechanics and Physics of Solids*, 47(2): 297-324, 1999.

[53] W.D. Nix, H. Gao. Indentation size effects in crystalline materials: A law for strain gradient plasticity. *Journal of the Mechanics and Physics of Solids*, 46(3): 411-425, 1998.

[54] H. Jiang, Y. Huang, Z. Zhuang, K.C. Hwang. Fracture in mechanism-based strain gradient plasticity. *Journal of the Mechanics and Physics of Solids*, 49(5): 979-993. 2001.

[55] M. Shi, Y. Huang, H. Jiang, K.C. Hwang, M. Li. The boundary-layer effect on the crack tip field in mechanism-based strain gradient plasticity. *International Journal of Fracture*, 112(1): 23-41, 2001.

[56] B. Liu, X. Qiu, Y. Huang, K.C. Hwang, M. Li, C. Liu. The size effect on void growth in ductile materials. *Journal of the Mechanics and Physics of Solids*, 51(7): 1171-1187, 2003.

[57] M.E. Gurtin, L. Anand. A theory of strain-gradient plasticity for isotropic, plastically irrotational materials. Part I: Small deformations. *Journal of the Mechanics and Physics of Solids*, 53(7): 1624-1649, 2005.

[58] M.E. Gurtin, L. Anand. A theory of strain-gradient plasticity for isotropic, plastically irrotational materials. Part II: Finite deformations. *International Journal of Plasticity*, 21(12): 2297-2318, 2005.

[59] P. Gudmundson. A unified treatment of strain gradient plasticity. *Journal of the Mechanics and Physics of Solids*, 52(6): 1379-1406, 2004.

[60] R. Venkatraman, J.C. Bravman. Separation of Film Thickness and Grain-Boundary Strengthening Effects in Al Thin-Films on Si. *Journal of Materials Research*, 7(8): 2040-2048, 1992.

[61] A. Acharya, J.L. Bassani. On non-local flow theories that preserve the classical structure of incremental boundary value problems. *IUTAM Symposium on Micromechanics of Plasticity and Damage of Multiphase Materials*, 46: 3-9420, 1996.

[62] M. Luo. Incompatibility theory of nonlocal plasticity and applications. PhD Thesis, University of Pennsylvania, 1998.

[63] H.H.M. Cleveringa, V.E. Giessen, A.Needleman. Comparison of discrete dislocation and continuum plasticity predictions for a composite material. Acta Materialia, 45: 3163-3179, 1997.

[64] Q. Ma, D.R. Clarke. Size dependent hardness of silver single crystal. *Journal of Materials Research*, 10: 853-863, 1995.

[65] S. Chen, T. Wang. Strain gradient theory with couple stress for crystalline solids. *European Journal of Mechanics - A/Solids*, 20(5): 739-756, 2001.

[66] K.W. McElhaney, J.J. Vlassak, W.D. Nix. Determination of indenter tip geometry and indentation contact area for depth-sensing indentation experiments. *Journal of Materials Research*, 13(5): 1300-1306, 1998.

[67] N.A. Stelmashenko, M.G. Walls, L.M. Brown, Y.V. Milman. Microindentations on W and Mo Oriented Single-Crystals - an Stm Study. *Acta Metallurgica Et Materialia*, 41(10): 2855-2865, 1993.

[68] W.J. Poole, M.F. Ashby, N.A. Fleck. Micro-hardness of annealed and work-hardened copper polycrystals. *Scripta Materialia*, 34(4): 559-564, 1996.

[69] G. Dehm, T. Wagner, T.J. Balk, E. Arzt, B.J. Inkson. Plasticity and interfacial dislocation mechanisms in epitaxial and polycrystalline Al films constrained by substrates. *Journal of Materials Science & Technology*, 18(2): 113-117, 2002.

[70] A. Misra, J.P. Hirth, R.G. Hoagland. Length-scale-dependent deformation mechanisms in incoherent metallic multilayered composites. *Acta Materialia*, 53(18): 4817-4824, 2005.

[71] R.G. Hoagland, R.J. Kurtz, C.H. Henager. Slip resistance of interfaces and the strength of metallic multilayer composites. *Scripta Materialia*, 50(6): 775-779, 2004.

[72] M.D. Uchic, D.M. Dimiduk, J.N. Florando, W.D. Nix. Sample dimensions influence strength and crystal plasticity. *Science*, 305(5686): 986-989, 2004.

[73] B. Ehrler, X.D. Hou, T.T. Zhu, K.M.Y. Png, C.J. Walker, A.J. Bushby, D.J. Dunstan. Grain size and sample size interact to determine strength in a soft metal. *Philosophical Magazine*, 88(25): 3043-3050, 2008.

[74] E.O. Hall. The Deformation and Ageing of Mild Steel .3. Discussion of Results. *Proceedings of the Physical Society of London Section B,*

64(381): 747-753, 1951.

[75] N.J. Petch. The Cleavage Strength of Polycrystals. *The Journal of the Iron and Steel Institute*, 174:25-28, 1953.

[76] J.D. Eshelby, F.C. Frank, F.R.N. Nabarro. The Equilibrium of Linear Arrays of Dislocations. *Philosophical Magazine*, 42(327): 351-364, 1951.

[77] J. Li, Y. Chou. The role of dislocations in the flow stress grain size relationships. *Metallurgical and Materials Transactions B*, 1(5): 1145-1159, 1970.

[78] N.F. Mott. Atomic Physics and the Strength of Metals. *Journal of the Institute of Metals*, 72(9): 367-380, 1946.

[79] J.C.M. Li. Direct Observation of Imperfections in Crystals. St. Louis, Missouri: Interscience (Wiley), New York, ed. J.B. Newkirk, J.H. Wernick, 1962.

[80] J.C.M. Li. Petch Relation and Grain Boundary Sources. *Transactions of the Metallurgical Society of Aime*, 227(1): 239-247, 1963.

[81] M.A. Meyers, E. Ashworth. A Model for the Effect of Grain-Size on the Yield Stress of Metals. *Philosophical Magazine a-Physics of Condensed Matter Structure Defects and Mechanical Properties*, 46(5): 737-759, 1982.

[82] J.R. Weertman. Mechanical Behavior of Nanocrystalline Metals. Nanostructured Materials: Processing, Properties and Applications, 2nd Edition, ed. Carl C. Koch, 398-399, 1992. ISBN: 0185514514.

[83] H. Conrad. Electron Microscopy and Strength of Crystals, ed. G. Thames and J. Washburn. New York: Interscience. 299–300, 1961.

[84] K.H. Chia, K. Jung, H. Conrad. Dislocation density model for the effect

of grain size on the flow stress of a Ti-15.2 at.% Mo beta-alloy at 4.2-650 K. *Materials Science and Engineering a-Structural Materials Properties Microstructure and Processing*, 409(1-2): 32-38, 2005.

[85] A.S.M. Aena. A study of flow characteristics of nanostructured Al-6082 alloy produced by ECAP under upsetting test. *Journal of Materials Processing Technology*, 209(2): 856-863, 2009.

[86] A.H. Chokshi, A. Rosen, J. Karch, H. Gleiter. On the Validity of the Hall-Petch Relationship in Nanocrystalline Materials. *Scripta Metallurgica*, 23(10): 1679-1683, 1989.

[87] M.A. Meyers, A. Mishra, D.J. Benson. Mechanical properties of nanocrystalline materials. *Progress in Materials Science*, 51(4): 427-556, 2006.

[88] J.R. Greer, J.T.M. De Hosson. Plasticity in small-sized metallic systems: Intrinsic versus extrinsic size effect. *Progress in Materials Science*, 56(6): 654-724, 2011.

[89] M. Ke, S.A. Hackney, W.W. Milligan, E.C. Aifantis. Observation and Measurement of Grain Rotation and Plastic Strain in Nanostructured Metal Thin-Films. *Nanostructured Materials*, 5(6): 689-697, 1995.

[90] Y.B. Wang, B.Q. Li, M.L. Sui, S.X. Mao. Deformation-induced grain rotation and growth in nanocrystalline Ni. *Applied Physics Letters*, 92(1): 011903, 2008.

[91] D. Farkas, S. Mohanty, J. Monk. Strain-driven grain boundary motion in nanocrystalline materials. *Materials Science and Engineering a-Structural Materials Properties Microstructure and Processing*, 493(1-2): 33-40, 2008.

[92] H. Van Swygenhoven, P.A. Derlet. Grain-boundary sliding in nanocrystalline fcc metals. *Physical Review B*, 64(22): 224105, 2001.

[93] H. Van Swygenhoven, P.M. Derlet, A. Hasnaoui. Atomic mechanism for dislocation emission from nanosized grain boundaries. *Physical Review B*, 66(2): 204101, 2002.

[94] V. Yamakov, D. Wolf, S.R. Phillpot, A.K. Mukherjee, H. Gleiter. Dislocation processes in the deformation of nanocrystalline aluminium by molecular-dynamics simulation. *Nature Materials*, 1(1): 45-48, 2002.

[95] V. Yamakov, D. Wolf, S.R. Phillpot, H. Gleiter. Grain-boundary diffusion creep in nanocrystalline palladium by molecular-dynamics simulation. *Acta Materialia*, 50(1): 61-73, 2002.

[96] W.M. Yin, S.H. Whang. The creep and fracture in nanostructured metals and alloys. *Journal of the Minerals Metals & Materials Society*, 57(1): 63-70. 2005.

[97] M. Legros, D.S. Gianola, K.J. Hemker. In situ TEM observations of fast grain-boundary motion in stressed nanocrystalline aluminum films. *Acta Materialia*, 56(14): 3380-3393, 2008.

[98] D.S. Gianola, S. Van Petegem, M. Legros, S. Brandstetter, H. Van Swygenhoven, K.J. Hemker. Stress-assisted discontinuous grain growth and its effect on the deformation behavior of nanocrystalline aluminum thin films. *Acta Materialia*, 54(8): 2253-2263, 2006.

[99] D.S. Gianola, C. Eberl, X.M. Cheng, K.J. Hemker. Stress-driven surface topography evolution in nanocrystalline Al thin films. *Advanced Materials*, 20(2): 303-308, 2008.

[100] D.S. Gianola, B. Mendis, X.M. Cheng, K.J. Hemker. Grain-size stabilization by impurities and effect on stress-coupled grain growth in nanocrystalline Al thin films. *Materials Science and Engineering A*, 483-484: 637-640, 2008.

[101] D.S. Gianola, D.H. Warner, J.F. Molinari, K.J. Hemker. Increased strain

rate sensitivity due to stress-coupled grain growth in nanocrystalline Al. *Scripta Materialia*, 55(7): 649-652, 2006.

[102] T.W. Clyne, P.J. Withers. An Introduction to Metal Matrix Composites. Cambridge University Press, 1995. ISBN 13: 9780521483575.

[103] J.J. Lewandowski, C. Liu, W.H. Hunt. Effects of Matrix Microstructure and Particle Distribution on Fracture of an Aluminum Metal Matrix Composite. *Materials Science and Engineering A*, 107: 241-255, 1989.

[104] E. Orowan. Internal Stress in metal and alloys. Symposium on Internal Stresses in Metals and Alloys, Session III discussion. London: Institue of Metals, 1948.

[105] R.W. Hertzberg. Deformation and Fracture mechanics of Engineering Materials. 4th Edition, John Wiley & Sons, New York, 142, 1996. ISBN: 0471012149, 9780471012146.

[106] D.J. Lloyd. Particle-Reinforced Aluminum and Magnesium Matrix Composites. *International Materials Reviews*, 39(1): 1-23, 1994.

[107] W.H. Hunt, J.R. Brockenbrough, P.E. Magnusen. An Al-Si-Mg Composite Model System - Microstructural Effects on Deformation and Damage Evolution. *Scripta Metallurgica Et Materialia*, 25(1): 15-20, 1991.

[108] Z. Ling, L. Luo, B. Dodd. Experimental-Study on the Formation of Shear Bands and Effect of Microstructure in Al-2124/Sic-P Composites under Dynamic Compression. *Journal De Physique IV*, 4(C8): 453-458, 1994.

[109] J.C. Fisher, E.W. Hart, R.H. Pry. The hardening of metal crystals by precipitate particles. *Acta Metallurgica*, 1(3): 336-339, 1953.

[110] H. Gao, Y. Huang. Geometrically necessary dislocation and

size-dependent plasticity. *Scripta Materialia*, 48(2): 113-118, 2003.

[111] F.C. Frank, J.H. van der Merwe, One-dimensional dislocations. I. Static theory. *Proceedings of the Royal Society A*, 198: 205-215, 1949.

[112] D.J. Dunstan, A.J. Bushby. Theory of deformation in small volumes of material. *Proceedings of the Royal Society of London Series A-Mathematical Physical and Engineering Sciences*, 460(2050): 2781-2796, 2004.

[113] P. Moreau, M. Raulic, K.M.Y. P'ng, G. Gannaway, P. Anderson, W.P. Gillin, A.J. Bushby, D.J. Dunstan. Measurement of the size effect in the yield strength of nickel foils. *Philosophical Magazine Letters*, 85(7): 339-343, 2005.

[114] C. Motz, O. Friedl, R. Pippan. Mechanical properties of micro-sized copper bending beams machined by the focused ion beam technique. *Acta Materialia,* 53: 4269-4279, 2005.

[115] W.C. Oliver, G.M. Pharr. An Improved Technique for Determining Hardness and Elastic-Modulus Using Load and Displacement Sensing Indentation Experiments. *Journal of Materials Research*, 7(6): 1564-1583, 1992.

[116] N. Gane, J.M. Cox. The micro-hardness of metals at very low loads. *Philosophical Magazine*, 22(179): 0881-0891, 1970.

[117] Y.Y. Lim, M.M. Chaudhri. The effect of the indenter load on the nanohardness of ductile metals: an experimental study on polycrystalline work-hardened and annealed oxygen-free copper. *Philosophical Magazine a-Physics of Condensed Matter Structure Defects and Mechanical Properties*, 79(12): 2979-3000, 1999.

[118] J.G. Swadener, E.P. George, G.M. Pharr. The correlation of the indentation size effect measured with indenters of various shapes.

Journal of the Mechanics and Physics of Solids, 50(4): 681-694, 2002.

[119] W.D. Nix. Mechanical properties of thin films. *Metallurgical and Materials Transactions A*, 20: 2217-2245, 1989.

[120] R. Dou, B. Derby. A universal scaling law for the strength of metalmicropillars and nanowires. *Scripta Materialia*, 61 (5): 524-527, 2009.

[121] J.R. Greer, W.C. Oliver, W.D. Nix. Size dependence of mechanical properties of gold at the micron scale in the absence of strain gradients. *Acta Materialia*, 53(6): 1821-1830, 2005.

[122] J.R. Greer, W.D. Nix. Nanoscale gold pillars strengthened through dislocation starvation. *Physical Review B*, 73(24): 245410, 2006.

[123] S.I. Rao, D.M. Dimiduk, T.A. Parthasarathy, M.D. Uchic, M. Tang, C. Woodward. Athermal mechanisms of size-dependent crystal flow gleaned from three-dimensional discrete dislocation simulations. *Acta Materialia*, 56(13): 3245-325, 2008.

[124] T.A. Parthasarathy, S.I. Rao, D.M. Dimiduk, M.D. Uchic, D.R. Trinkle. Contribution to size effect of yield strength from the stochastics of dislocation source lengths in finite samples. *Scripta Materialia*, 56(4): 313-316, 2007.

[125] Rao, S.I. D.M. Dimiduk, M. Tang, T.A. Parthasarathy, M.D. Uchic, C. Woodward. Estimating the strength of single-ended dislocation sources in micron-sized single crystals. *Philosophical Magazine*, 87(30): 4777-4794, 2007.

[126] M.D. Uchic, D.M. Dimiduk, R. Wheeler, P.A. Shade, H.L. Fraser. Application of micro-sample testing to study fundamental aspects of plastic flow. *Scripta Materialia*, 54(5): 759-764, 2006.

[127] J.W. Matthews, A.E. Blakeslee. Defects in epitaxial multilayers. I. Misfit dislocations. *Journal of Crystal Growth*, 27: 118-125, 1974.

[128] J.W. Matthews, A.E. Blakeslee. Defects in epitaxial multilayers. II. Dislocation pile-ups, threading dislocations, slip lines, and cracks. *Journal of Crystal Growth*, 29: 273-280, 1975.

[129] L.B. Freund. A criterion for arrest of a threading dislocation in a strained epitaxial layer due to an interface misfit dislocation in its path. *Journal of Applied Physics*, 68: 2073-2080, 1990.

[130] O. Kraft, P.A. Gruber, R. Mönig, D. Weygand. *Plasticity in Confined Dimensions. Annual Review of Materials Research*, 40: 293-317, 2010.

[131] H. Gao, L. Zhang, S.P. Bakerc. Dislocation core spreading at interfaces between metal films and amorphous substrates. *Journal of the Mechanics and Physics of Solids*, 50: 2169-2202, 2002.

[132] R.W. Armstrong. On Size Effects in Polycrystal Plasticity. *Journal of the Mechanics and Physics of Solids*, 9(3): 196-199, 1961.

[133] R.M. Keller, S.P. Baker, E. Arzt. Quantitative analysis of strengthening mechanisms in thin Cu films: Effects of film thickness, grain size, and passivation. *Journal of Materials Research*, 13(5): 1307-1317, 1998.

[134] X.D. Hou, A.J. Bushby, N.M. Jennett. Study of the interaction between the indentation size effect and Hall-Petch effect with spherical indenters on annealed polycrystalline copper. *Journal of Physics D-Applied Physics*, 41(7): 074006, 2008.

[135] X.X. Chen, A.H.W. Ngan. Specimen size and grain size effects on tensile strength of Ag microwires. *Scripta Materialia*, 64(8): 717-720, 2011.

[136] G. Khatibi, R. Srickler, V. Gröger, B. Weiss. Tensile properties of thin

Cu-wires with a bamboo microstructure. *Journal of Alloys and Compounds*, 378(1-2): 326-328, 2004.

[137] I. Baker. Recovery, recrystallization and grain growth in ordered alloys. *Intermetallics*, 8(9-11): 1183-1196, 2000.

[138] R.W. Cahn, P. Haasen, eds. Physical Metallugry 3rd revised and enlarged edition. Recovery and recrystallization, ed. E.W. Cahn. North-Holland physics publishing: Amsterdam, New York, 1983.

[139] R.D. Doherty, D.A. Hughes, F.J. Humphreys, J.J. Jonas, D.J. jensen, M.E. Kassner, W.E. King, T.R. McNelley, H.J. McQueen, A.D. Rollett. Current issues in recrystallization: a review. *Materials Science and Engineering a-Structural Materials Properties Microstructure and Processing,* 238(2): 219-274, 1997.

[140] G.J. Qi, S. Zhang. Recrystallization of gold alloys for producing fine bonding wires. *Journal of Materials Processing Technology,* 68(3): 288-293, 1997.

[141] F.J. Humphreys, M. Hatherly. Recrystallization and Related Annealing Phenomena, Elsevier Ltd., Oxford: Pergamon Press, 1995.

[142] R.D. Doherty, I. Samajdar, C.T. Necker, H.E. Vatne, E. Nes. Microstructural and Crystallographic Aspects of Recrystallization. Proceedings of the 16th Riso International Symposium on Materials Science, ed. N. Hansen. Denmark: Riso National Lab, Roskilde, 1995.

[143] B. Duggan, C.Y. Chung. Effect of cube nucleus distribution on cube texture. Mater. Sci. Forum, ICOTOM 10, ed. H.J. Bunge, 157-163, 1994.

[144] T. Mukai, S. Suresh, K. Kita, H. Sasaki, N. Kobayashi, K. Higashi, A. Inoue. Nanostructured Al-Fe alloys produced by e-beam deposition: static and dynamic tensile properties. *Acta Materialia*, 51(14): 4197-4208, 2003.

[145] R.D. Emery, G.L. Povirk. Tensile behavior of free-standing gold films. Part I. Coarse-grained films. *Acta Materialia*, 51(7): 2067-2078, 2003.

[146] Y.H. Chew, C.C. Wong, F. Wulff, F.C. Lim, H.M. Goh. Strain rate sensitivity and Hall-Petch behavior of ultrafine-grained gold wires. *Thin Solid Films*, 516(16): 5376-5380, 2008.

[147] R. Schwaiger, B. Moser, M. Dao, N. Chollacoop, S. Suresh. Some critical experiments on the strain-rate sensitivity of nanocrystalline nickel. *Acta Materialia*, 51(17): 5159-5172, 2003.

[148] T. Takasugi, T. Tsuyumu, Y. Kaneno, H. Inoue. Anomalous tensile elongation increase of moisture-embrittled Co3Ti alloys in the low strain rate range. *Scripta Materialia*, 43(5): 397-402, 2000.

[149] X.M. Cao, Q.Q. Duan, X.W. Li. Anomalous Effect of Strain Rate on the Tensile Elongation of Coarse-Grained Pure Iron with Grain Boundary Micro-Voids. *Advanced Materials Research*, 214: 334-338, 2011.

[150] J.H. Xu, T. Leonhardt, J. Farrell, M. Effgen, T.G. Zhai. Anomalous strain-rate effect on plasticity of a Mo-Re alloy at room temperature. *Materials Science and Engineering a-Structural Materials Properties Microstructure and Processing*, 479(1-2): 76-82, 2008.

[151] K. Matsubara, Y. Miyahara, Z. Horita, T.G. Langdon. Developing Superplasticity in a Magnesium Alloy Through a Combination of Extrusion and ECAP. *Acta Materialia*, 51: 3073-3084, 2003.

[152] T. Tanaka, M. Kohzu, Y. Takigawa, K. Higashi. Low cycle fatigue behavior of Zn–22mass%Al alloy exhibiting high-strain-rate super-plasticity at room temperature. *Scripta Materialia*, 52(3): 231-236, 2005.

[153] T.G. Nieh, C.H. Henshall, J. Wadsworth. Creep-Rupture of a Silicon-Carbide Reinforced Aluminum Composite. *Scripta Metallurgica et Materialia*, 18: 1405-1508, 1984.

[154] T.G. Langdon. The Mechanical Properties of Superplastic Materials. *Metallurgical and Materials Transactions A*, 1 3A: 689-701, 1982.

[155] M.W. Phaneuf. Applications of focusedionbeammicroscopy to materials science specimens. *Micron*, 30(3): 277-288, 1999.

[156] ASTME 112-10 - Standard Test Methods for Determining Average Grain Size, 9-10, 2010.

[157] M. Calcagnotto, D. Ponge, E. Demir, D. Raabe. Orientation gradients and geometrically necessary dislocations in ultrafine grained dual-phase steels studied by 2D and 3D EBSD. *Materials Science and Engineering A*, 527: 2738-2746, 2010.

[158] L.P. Kubin, A. Mortensen. Geometrically necessary dislocations and strain-gradient plasticity: a few critical issues. *Scripta Materialia*, 48(2): 119-125, 2003.

[159] E. Demir, D. Raabe. Mechanical and microstructural single-crystal Bauschinger effects: Observation of reversible plasticity in copper during bending. *Acta Materialia*, 58(18): 6055-6063, 2010.

[160] M. Walter, O. Kraft. A new method to measure torsion moments on small-scaled specimens. *Review of Scientific Instruments*, 82(3): 035109, 2011.

[161] M. Walter, O. Kraft, M. Klotz. Vorrichtung zur Bestimmung von Torsionsmomenten im Submikronewtonmeterbereich, Karlsruher Institut für Technologie, European patent No. 1903326, 2011.

[162] P.E. Viljoen, E.S. Lambers, P.H. Holloway. Reaction between Diamond and Titanium for Ohmic Contact and Metallization Adhesion Layers. *Journal of Vacuum Science & Technology B*, 12(5): 2997-3005, 1994.

[163] C.G. Fountzoulas, D.M. Potrepka, S.C. Tidrow. Microstructural and Electrical Characterization of Barium Strontium Titanate-based Solid Solution Thin Films Deposited on Ceramic Substrates by Pulsed Laser Deposition. *Materials Research Society Symposium Proceedings*, 720: 29-34, 2002.

[164] S. Levchuk, A. Brendel, S. Lindig, H. Bolt. Investigation of the interface reactions between silicon carbide fibres and 9Cr steel matrix. 2007.

[165] W. Ramberg, W. R. Osgood. Description of stress-strain curves by three parameters, in National Advisory Committee For Aeronautics. Technical Note No. 902: Washington DC, 1943.

[166] C. Keller, E. Hug, X. Feaugas. Microstructural size effects on mechanical properties of high purity nickel. *International Journal of Plasticity*, 27(4): 635-654, 2011.

[167] P.J.M. Janssen, T.H. de Keijser, M.G.D. Geers. An experimental assessment of grain size effects in the uniaxial straining of thin Al sheet with a few grains across the thickness. *Materials Science and Engineering: A*, 419: 238-248, 2006.

Appendix 1

The Young's modulus of gold wires annealed at 400°C for 2 hours determined from tensile tests with unloading segments

Diameter	$E0$ [GPa]	$E1$ [GPa]	$E2$ [GPa]	$E3$ [GPa]	$E4$ [GPa]	$E5$ [GPa]	$E6$ [GPa]	Average value [GPa]
15 μm	59	67	66	68	66	66	71	66
	66	66	65	72	70	65	62	
	54	66	67	66	61	61	54	
25 μm	64	74	74	74	73	73	73	72
	62	71	72	71	72	72	72	
	60	72	71	71	72	71	71	
40 μm	65	70	70	72	73	74	73	72
	69	69	72	73	74	73	73	
	64	70	74	72	73	73	73	
60 μm	64	68	70	71	71	70	72	71
	66	71	71	72	72	72	72	
	70	72	72	72	73	73	71	

Appendix 2

(a): The Young's modulus of gold wires in the fully recrystallized state determined from tensile tests with unloading segments

Diameter	$E0$ [GPa]	$E1$ [GPa]	$E2$ [GPa]	$E3$ [GPa]	$E4$ [GPa]	$E5$ [GPa]	$E6$ [GPa]	Average value [GPa]
15 μm	32	46	47	51	52	53	51	51
	37	54	54	55	55	54	55	
	51	50	49	49	50	50	48	
25 μm	56	62	62	63	62	64	64	64
	30	62	63	64	63	65	64	
	77	64	65	65	66	66	66	
40 μm	42	73	75	76	75	75	75	75
	38	75	75	75	76	76	76	
	54	75	74	75	74	75	75	
60 μm	38	72	72	74	72	74	71	73
	54	73	73	74	73	75	75	
	55	72	72	73	72	74	74	

(b): The Young's modulus of 12.5 and 17.5 µm gold wires in the fully recrystallized state determined from tensile tests with unloading segments

Diameter	E0 [GPa]	E1 [GPa]	E2 [GPa]	E3 [GPa]	E4 [GPa]	E5 [GPa]	E6 [GPa]	Average value [GPa]
12.5 µm	49	54	64	52	77	57		65
	46	100	63	20	31	58		
	52	126	30	68	89	88		
17.5 µm	117	67	67	68	66	68	73	64
	56	62	65	69	55	60	61	
	149	65	56	61	78	71	51	

Schriftenreihe
des Instituts für Angewandte Materialien

ISSN 2192-9963

Die Bände sind unter www.ksp.kit.edu als PDF frei verfügbar oder
als Druckausgabe bestellbar.

Band 1 Prachai Norajitra
    Divertor Development for a Future Fusion Power Plant. 2011
    ISBN 978-3-86644-738-7

Band 2 Jürgen Prokop
    Entwicklung von Spritzgießsonderverfahren zur Herstellung
    von Mikrobauteilen durch galvanische Replikation. 2011
    ISBN 978-3-86644-755-4

Band 3 Theo Fett
    New contributions to R-curves and bridging stresses –
    Applications of weight functions. 2012
    ISBN 978-3-86644-836-0

Band 4 Jérôme Acker
    Einfluss des Alkali/Niob-Verhältnisses und der Kupferdotierung
    auf das Sinterverhalten, die Strukturbildung und die Mikro-
    struktur von bleifreier Piezokeramik $(K_{0,5}Na_{0,5})NbO_3$. 2012
    ISBN 978-3-86644-867-4

Band 5 Holger Schwaab
    Nichtlineare Modellierung von Ferroelektrika unter
    Berücksichtigung der elektrischen Leitfähigkeit. 2012
    ISBN 978-3-86644-869-8

Band 6 Christian Dethloff
    Modeling of Helium Bubble Nucleation and Growth
    in Neutron Irradiated RAFM Steels. 2012
    ISBN 978-3-86644-901-5

Band 7 Jens Reiser
    Duktilisierung von Wolfram. Synthese, Analyse und Charak-
    terisierung von Wolframlaminaten aus Wolframfolie. 2012
    ISBN 978-3-86644-902-2

Band 8 Andreas Sedlmayr
    Experimental Investigations of Deformation Pathways
    in Nanowires. 2012
    ISBN 978-3-86644-905-3

Band 19   Christian Mennerich
**Phase-field modeling of multi-domain evolution in ferromagnetic
shape memory alloys and of polycrystalline thin film growth.** 2013
ISBN 978-3-7315-0009-4

Band 20   Spyridon Korres
**On-Line Topographic Measurements of Lubricated Metallic Sliding
Surfaces.** 2013
ISBN 978-3-7315-0017-9

Band 21   Abhik Narayan Choudhury
**Quantitative phase-field model for phase transformations in
multi-component alloys.** 2013
ISBN 978-3-7315-0020-9

Band 22   Oliver Ulrich
**Isothermes und thermisch-mechanisches Ermüdungsverhalten von
Verbundwerkstoffen mit Durchdringungsgefüge (Preform-MMCs).** 2013
ISBN 978-3-7315-0024-7

Band 23   Sofie Burger
**High Cycle Fatigue of Al and Cu Thin Films by a Novel High-Throughput
Method.** 2013
ISBN 978-3-7315-0025-4

Band 24   Michael Teutsch
**Entwicklung von elektrochemisch abgeschiedenem LIGA-Ni-Al für
Hochtemperatur-MEMS-Anwendungen.** 2013
ISBN 978-3-7315-0026-1

Band 25   Wolfgang Rheinheimer
Zur Grenzflächenanisotropie von SrTiO$_3$. 2013
ISBN 978-3-7315-0027-8

Band 26   Ying Chen
**Deformation Behavior of Thin Metallic Wires under Tensile
and Torsional Loadings.** 2013
ISBN 978-3-7315-0049-0